Collins

ROYAL
OBSERVATORY
GREENWICH

Night Sky
ALMANAC

A STARGAZER'S
GUIDE TO

2021

Storm Dunlop & Wil Tirion

Collins
of HarperCollins Publishers
a Road
briggs
ow G64 2QT
v.harpercollins.co.uk

.n association with
Royal Museums Greenwich, the group name for the National Maritime Museum,
Royal Observatory Greenwich, Queen's House and *Cutty Sark*
www.rmg.co.uk

© HarperCollins Publishers 2020
Text and illustrations © Storm Dunlop and Wil Tirion
Cover illustrations © Julia Murray
Images and illustrations see acknowledgements page 252

The contents of this publication are believed correct at the time of printing.
Nevertheless the publisher can accept no responsibility for errors or omissions,
changes in the detail given or for any expense or loss thereby caused.

HarperCollins does not warrant that any website mentioned in this title will
be provided uninterrupted, that any website will be error free, that defects will
be corrected, or that the website or the server that makes it available are free
of viruses or bugs. For full terms and conditions please refer to the site
terms provided on the website.

A catalogue record for this book is available from the British Library

ISBN 978-0-00-840360-7

10 9 8 7 6 5 4 3 2

Printed and bound by CPI Group (UK) Ltd,
Croydon CR0 4YY

If you would like to comment on any aspect of this book,
please contact us at the above address or online.
e-mail: collins.reference@harpercollins.co.uk

 facebook.com/CollinsAstronomy
 @CollinsAstro

MIX
Paper from
responsible sources
FSC™ C007454
www.fsc.org

This book is produced from independently certified
FSC™ paper to ensure responsible forest management.

For more information visit: www.harpercollins.co.uk/green

Contents

Foreword

Foreword

For thousands of generations, your ancestors have marvelled at the night sky above them. They pondered its meaning and distance, tracing patterns and telling stories with the stars; they watched with great interest as the Moon and a handful of bright, wandering lights traversed the sky independently; sometimes they looked on in awe as a new visitor drifted above, with a long tail in tow; at other times they stayed up, alert, as stars appeared to streak across the sky with alarming speed; perhaps they cowered in fear when, on rare occasions, the Moon turned blood red or challenged the daylight by taking a bite from the Sun.

Gradually, superstition gave way to understanding. Comprehension of the night sky's cyclical nature proved foundational to an inevitable agricultural revolution, which propelled the progress of civilisation forward, forever cementing astronomy's influence on our lives. The stars have also permeated into our wider cultural heritage, appearing in religious texts and great works of art – visual, poetic and musical. The Moon, our unfailing celestial companion, altered the course of life on Earth through its mastery of the tides, and has since become a frontier of exploration for our species – a stepping stone towards our spacefaring future. It is remarkable how far we have come since humans first turned their attention to the stars over 50,000 years ago.

Today, our lives are very different from theirs, but the curiosity driving the earliest stargazers has not left us. It has only become muffled by the increasing intrusion of artificial light pollution that has brightened our skies. The stars, planets and Moon still play overhead as they always have, and each year many annual meteor showers and other events go underappreciated as fewer of us enjoy regular access to the night sky. Yet, with only a little guidance you can rekindle this deeply ingrained fascination and, armed with the extensive knowledge of the modern astronomical age, explore the universe we call home.

This Almanac invites you to reconnect with your ancestral stargazing roots, by following the progress of the constellations throughout the seasons. With the aid of its monthly calendars and maps, you will chart the rhythm of the lunar phases, discover events that light up the sky for brief periods, and explore the rich tapestry of characters that adorn the starry canvas overhead. You can delve as deeply as you like, or follow your own favourite subject throughout the year. The treasures of the night sky will be waiting for you.

Tom Kerss

Introduction

The aim of this book is to help people to find their way around the night sky and to understand what is visible every month, from anywhere in the world. The stars that may be seen depend on where you are on Earth, but even if you travel widely, this book will show you what you can see. The night sky also changes from month to month and these changes, together with some of the significant events that occur during the year are described and illustrated.

The charts that are used differ considerably from those found in most astronomy books, and have been specifically designed for use anywhere in the world. A full description of how to use and understand the monthly charts is given on pages 32 to 35.

Sunrise, sunset and twilight

The conditions for observing naturally vary over the course of the year and one's location on Earth. Sunrise and sunset vary considerably, depending in particular on one's latitude. Sunrise and sunset times are given each month for nine different locations around the world. These places are shown in a **bold** typeface on the world map on pages 36 and 37. Sunrise and sunset times are given for the first and last days in every month, for these specific locations. Another factor that influences what may be seen is twilight at dusk and dawn. Again, this varies considerably with one's latitude on Earth. The diagrams on pages 238 to 241 show how this varies for the nine locations, which have been chosen to show the range of variation, rather than just for the importance of the places that have been included. The different stages of twilight and how they affect observing are also explained there.

Moonlight

Yet another factor that affects the visibility of objects is the amount of moonlight in the sky. At Full Moon, it may be very difficult to see some of the fainter stars and objects, and even when the Moon is at a smaller phase it seriously interferes with visibility if it is near the stars or planets in which you are interested. A full lunar calendar is given for each month and may be used to see when nights are likely to be darkest and best for observing.

The celestial sphere

All the objects in the sky (including the Sun, Moon, and stars) appear to lie at some indeterminate distance on a large sphere, centred on the Earth. This celestial sphere has various reference points and features that are related to those of the Earth. If the Earth's rotational axis is extended, for example, it points to the North and South Celestial Poles, which are thus in line with the North and South Poles on Earth. As shown in the diagrams, the altitude of the celestial pole is equal to the observer's latitude, whether in the north or south. Similarly, the celestial equator lies in the same plane as the Earth's equator, and divides the sky into northern and southern hemispheres.

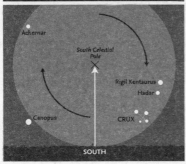

It is useful to know some of the special terms for various parts of the sky. As seen by an observer, half of the celestial sphere is invisible, below the horizon. The point directly overhead is known as the **zenith**, and this point is shown on the monthly charts for several different latitudes, where it is an important reference point. The (invisible) point below one's feet is the **nadir**.

The altitude of the Celestial Pole equals the observer's latitude.

The line running from the north point on the horizon, up through the zenith and then down to the south point is the **meridian**. This is an important invisible line in the sky, because objects are highest in the sky, and thus easiest to see, when they cross the meridian in the south. Objects are said to transit, when they cross this line in the sky.

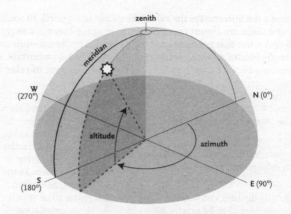

Measuring altitude and azimuth on the celestial sphere.

In this book, reference is sometimes made in the text and in the diagrams to the standard compass points around the horizon. The position of any object in the sky may be described by its **altitude** (measured in degrees above the horizon) and its **azimuth** (measured in degrees from north, 0°, through east, 90°, south, 180°, and west, 270°). Experienced amateurs and professional astronomers also use another system of specifying locations on the celestial sphere, but that need not concern us here, where the simpler method will suffice.

The celestial sphere appears to rotate about an invisible axis, running between the north and south celestial poles. The location (i.e., the altitude) of the celestial poles depends entirely on the observer's position on Earth or, more specifically, their latitude.

The ecliptic and the zodiac
Another important line on the celestial sphere is the Sun's apparent path against the background stars – in reality the result of the Earth's orbit around the Sun. This is known as the **ecliptic.** The point where the Sun, apparently moving along the ecliptic, crosses the celestial equator from south to north is known as the vernal (or northern spring) equinox, which occurs around March 21. At this time (and at the northern autumnal equinox, on September 22 or

23, when the Sun crosses the celestial equator from north to south) day and night are almost exactly equal in length. (There is a slight difference, but that need not concern us here.) The vernal equinox is currently located in the constellation of Pisces, and is important in astronomy because it defines the zero point for a system of celestial coordinates, which is, however, not used in this book.

The Moon and planets are to be found in a band of sky that extends 8° on either side of the ecliptic. This is because the orbits of the Moon and planets are inclined at various angles to the ecliptic (i.e., to the plane of the Earth's orbit). This band of sky is known as the zodiac, and when originally devised, consisted of twelve *constellations*, all of which were considered to be exactly 30° wide. When the constellation boundaries were formally established by the International Astronomical Union in 1930, the exact extent of most constellations was altered, and nowadays, the ecliptic passes through thirteen constellations. Because of the boundary changes, the Moon and planets may actually pass through several other constellations that are adjacent to the original twelve.

> The smallest constellation is *Crux*. It covers an area of just 68 square degrees.

The constellations

In the western astronomical tradition, the celestial sphere has always been divided into various constellations, most dating back to antiquity and usually associated with certain myths or legendary people and animals. Nowadays, 88 constellations cover the whole sky, and their boundaries have been fixed by international agreement. Their names (in Latin) are largely derived from Greek or Roman originals. (A full list of the constellations is given on pages 244 to 246, with their names in English, their abbreviations, and genitive forms.) Some of the names of the most prominent stars are of Greek or Roman origin, but many are derived from Arabic names. Some bright stars have no individual names, and

> The largest constellation is *Hydra*. It covers an area of 1303 square degrees, and more than 7 hours of Right Ascension (105°).

for many years, they were identified by terms such as 'the star in Hercules' right foot'. A more sensible scheme was introduced by the German astronomer Johannes Bayer in the early seventeenth century. Following his scheme – which is still used today – most of the brightest stars are identified by a Greek letter followed by the genitive form of the constellation's Latin name. An example is the Pole Star, also known as Polaris and α Ursae Minoris. The Greek alphabet is shown on page 246.

Asterisms

Apart from the constellations, certain groups of stars, which may form a small part of a larger constellation, are readily recognizable and have been given individual names. These groups are known as asterisms, and the most famous (and well-known) is the 'Plough' or 'Big Dipper', the common name for the seven brightest stars in the constellation of Ursa Major, the Great Bear. The names and identifications of some popular asterisms are given in the list on page 247.

The Moon

As it passes across the sky from west to east in its orbit around the Earth, the Moon moves by approximately its diameter (about half a degree) in an hour. Normally, in its orbit, the Moon passes above or below the direct line between Earth and Sun (at New Moon) or outside the area obscured by the Earth's shadow (at Full Moon). Occasionally, however, the three bodies are more-or-less perfectly aligned to give an eclipse: a solar eclipse at New Moon, or a lunar eclipse at Full Moon. Depending on the exact circumstances, a solar eclipse may be merely partial (when the Moon does not cover the whole of the Sun's disc); annular (when the Moon is too far from Earth in its orbit to appear large enough to hide the whole of the Sun); or total. Total and annular eclipses are visible from very restricted areas of the Earth, but partial eclipses are normally visible over a wider area. Two forms of solar eclipse occur this year, and are described in detail in the appropriate month.

Precautions must always be taken when viewing even partial phases of a solar eclipse to avoid damage to your eyes. Only ever use proper eclipse glasses, or a proper solar filter over the full objective of a telescope. The glass 'solar filters' sometimes provided with cheap telescopes should never be used. They are unsafe.

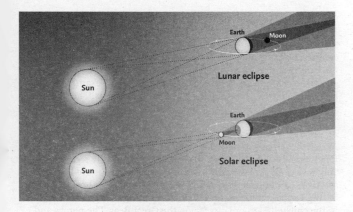

When the Moon passes through the Earth's shadow (top), a lunar eclipse occurs. When it passes in front of the Sun (below) a solar eclipse occurs.

Somewhat similarly, at a lunar eclipse, the Moon may pass through the outer zone of the Earth's shadow, the penumbra (in a penumbral eclipse, which is not generally perceptible to the naked eye); pass so that just part of the Moon is within the darkest part of the Earth's shadow, the umbra (in a partial eclipse); or completely within the umbra (in a total eclipse). Unlike solar eclipses, lunar eclipses are visible from large areas of the Earth. Again, these are described in detail in the relevant month.

Occasionally, as it moves across the sky, the Moon passes between the Earth and individual planets or distant stars, giving rise to an *occultation*. As with solar eclipses, such occultations are visible from restricted areas of the world, but certain significant occultations are described in detail.

The Planets
Because the planets are always moving against the background stars, they are treated in some detail in the monthly pages and information is given when they are close to other planets, the Moon, or any of five bright stars that lie near the ecliptic. Such events are known as *appulses* or, more frequently, as *conjunctions*. (There are technical differences in the way these terms are defined – and

should be used – in astronomy, but these need not concern us here.) The term conjunction is also used when a planet is either directly behind or in front of the Sun, as seen from Earth. (Under normal circumstances it will then be invisible.) The conditions of most favourable visibility depend on whether the planet is one of the two known as *inferior planets* (Mercury and Venus) or one of the three *superior planets* (Mars, Jupiter and Saturn) that are covered in detail. Brief details of the fainter superior planets, Uranus and Neptune, are given, especially when they come to opposition.

The inferior planets are most readily seen at eastern or western *elongation*, when their angular distance from the Sun is greatest. For superior planets and minor planets, they are best seen at *opposition*, when they are directly opposite the Sun in the sky, and cross the meridian at local midnight.

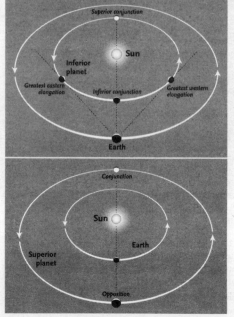

Inferior planet.

Superior planet.

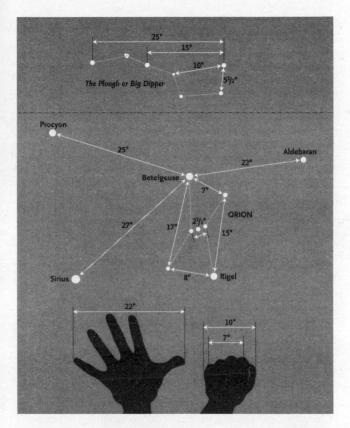

Measuring angles in the sky.

It is often useful to be able to estimate angles on the sky, and approximate values may be obtained by holding one hand at arm's length. The various angles are shown in the diagram, together with the separations of the various stars in the asterism, known as the Plough or Big Dipper, and also for stars around the constellation of Orion.

Events

A number of interesting events are shown in diagrams for each month. They involve the planets and the Moon, sometimes showing them in relation to specific stars. Events have been chosen as they will appear from one of three different locations: from London; from the central region of the USA; or from Sydney in Australia. Naturally, these events are visible from other locations, but the appearance of the objects on the sky will differ slightly from the diagrams. A list of major astronomical events in 2021 is given on pages 19 and 20.

Meteors

At some time or other, nearly everyone has seen a *meteor* – a 'shooting star' – as it flashed across the sky. The particles that cause meteors – known technically as 'meteoroids' – range in size from that of grain of sand (or even smaller) to the size of a pea. On any night of the year

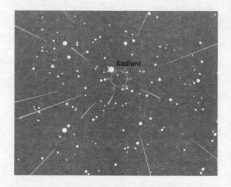

Meteor shower (showing the April Lyrid radiant).

there are occasional meteors, known as *sporadics*, that may travel in any direction. These occur at a rate that is normally between 3 and 8 in an hour. Far more important, however, are *meteor showers*, which occur at fixed periods of the year, when the Earth encounters a trail of particles left behind by a comet or, very occasionally, by a minor planet (asteroid). Meteors always appear to diverge from a single point on the sky, known as the *radiant*, and the radiants of major showers are shown on the charts.

Meteors that come from a circular area, 8° in diameter, around the radiant are classed as belonging to the particular shower. All others that do not come from that area are sporadics (or, occasionally, from another shower that is active at the same time). A list of the major meteor showers is given on the next page.

Although the positions of the various shower radiants are shown on the charts, looking directly at the radiant is not the most effective way of seeing meteors. They are most likely to be noticed if one is looking about 40–45° away from the radiant position. (This is approximately two hand-spans as shown in the diagram for measuring angles on page 13.)

Shower	Dates of activity 2021	Date of maximum 2021	Possible hourly rate
Quadrantids	Dec. 28 to Jan. 12	Jan. 3–4	120
α-Centaurids	Jan. 31 to Feb. 20	Feb. 8	6
γ-Normids	Feb. 24 to Mar. 24	Mar. 14–15	6
April Lyrids	Apr. 13–29	Apr. 22	18
π-Puppids	Apr. 14 to Apr. 27	Apr. 23–24	var.
η-Aquariids	Apr. 18 to May 27	May 6	40
Piscis Austrinids	Jul. 14 to Aug. 27	Jul. 28	5
α-Capricornids	Jul. 2 to Aug. 14	Jul. 30	5
Southern δ-Aquariids	Jul. 13 to Aug. 24	Jul. 30	< 25
Perseids	Jul. 16 to Aug. 23	Aug. 12–13	150
α-Aurigids	Aug. 28 to Sep. 5	Sep. 1	6
Southern Taurids	Sep. 10 to Nov. 20	Oct. 10	< 5
Draconids	Oct. 7–11	Oct. 8–9	var.
Orionids	Oct. 1 to Nov. 6	Oct. 21	25
Northern Taurids	Oct. 20 to Dec. 10	Nov. 12	< 5
Leonids	Nov. 5–29	Nov. 17–18	< 15
Phoenicids	Nov. 22 to Dec. 9	Dec. 2	var.
Puppid Velids	Nov. 30 to Dec. 14	Dec. 7	10
Geminids	Dec. 3–16	Dec. 14	120+
Ursids	Dec. 17–26	Dec. 22–23	< 10

Other objects

Certain other objects may be seen with the naked eye under good conditions. Some were given names in antiquity – Praesepe is one example – but many are known by what are called 'Messier numbers', the numbers in a catalogue of nebulous objects compiled by Charles Messier in the late-eighteenth century. Some, such as the Andromeda Galaxy, M31, and the Orion Nebula, M42, may be seen faintly by the naked eye, but all those given in the list here will benefit from the use of binoculars.

Apart from galaxies, such as M31, which contain thousands of millions of stars, there are also two types of cluster: open clusters, such as M45, the Pleiades, which may consist of a few dozen to some hundreds of stars; and globular clusters, such as M13 in Hercules, which are spherical concentrations of many thousands of stars.

In 1781, **Charles Messier** (26 June 1730 – 12 April 1817) published the final version of his catalogue of 110 nebulous objects and faint star clusters that might be confused with comets. The objects in this catalogue are still known as the Messier objects and are always quoted as 'M' numbers.

One or two gaseous nebulae, consisting of gas illuminated by stars within them are also visible. The Orion Nebula, M42, is one, and is illuminated by the group of four stars, known as the Trapezium, which may be seen within it by using a good pair of binoculars. A list of interesting objects is given on the next page.

Eta Carinae (η Carinae) is one of the most massive and luminous stars known. It is estimated to have a mass between 120 and 150 times that of the Sun, and be between four and five million times as luminous.

Dates and time

Astronomers, worldwide, use a standardized method of expressing the date and time. This prevents confusion in comparing observations made by different observers. The various elements are given in descending order: year, month (three-letter abbreviation to prevent confusion over using numbers), day, hour, minutes,

Some Interesting Objects

Messier IC / NGC	Name	Type	Constellation
—	47 Tucanae	globular cluster	Tucana
—	Hyades	open cluster	Taurus
—	Double Cluster	open cluster	Perseus
—	Melotte 111	open cluster	Coma Berenices
M3	—	globular cluster	Canes Venatici
M4	—	globular cluster	Scorpius
M8	Lagoon Nebula	gaseous nebula	Sagittarius
M11	Wild Duck Cluster	open cluster	Scutum
M13	Hercules Cluster	globular cluster	Hercules
M15	—	globular cluster	Pegasus
M22	—	globular cluster	Sagittarius
M27	Dumbbell Nebula	planetary nebula	Vulpecula
M31	Andromeda Galaxy	galaxy	Andromeda
M35	—	open cluster	Gemini
M42	Orion Nebula	gaseous nebula	Orion
M44	Praesepe	open cluster	Cancer
M45	Pleiades	open cluster	Taurus
M57	Ring Nebula	planetary nebula	Lyra
M67	—	open cluster	Cancer
IC 2602	Southern Pleiades	open cluster	Carina
NGC 752	—	open cluster	Andromeda
NGC 3242	Ghost of Jupiter	planetary nebula	Hydra
NGC 3372	Eta Carinae Nebula	gaseous nebula	Carina
NGC 4755	Jewel Box	open cluster	Crux
NGC 5139	Omega Centauri	globular cluster	Centaurus

seconds. (In extreme cases, fractions of minutes or seconds may be used.) The date and time are that on the Greenwich meridian (GMT), and ignore any changes for Summer Time / Daylight Saving Time (DST), and any adjustments for local time at the observer's location. This standard is known as Coordinated Universal Time (UTC). In this book and in many others this is generally given as Universal Time (UT). All times given in this book are in UT.

To avoid problems over the changes involved in moving to and from Summer Time / Daylight Saving Time (and the complications over the beginning and end dates) and also the adjustments for local time, experienced astronomers set a (cheap) watch or clock to Universal Time and keep it that way. Smartphone users may use a simple world clock app, provided they lock it to the time (GMT) on the Greenwich Meridian.

Similarly, the date given for an event is the date as it applies at the Greenwich meridian, i.e., in UT. Occasionally this may differ from the date as given by your local time. An event that occurs (say) late in the night in Europe may seem to occur on the previous day to an observer to the west (such as in the USA), when local time is taken into account. This is another complication that is avoided by using the Universal Time standard.

In 1801, the Italian astronomer Giuseppe Piazzi discovered what appeared to be a new planet orbiting between Mars and Jupiter, and named it Ceres. William Herschel proved it to be a very small object, calculating it to be only 320 km in diameter, and not a planet. He proposed the name asteroid, and soon other similar bodies were found. We now know that Ceres is 952 km in diameter. It is now considered to be a dwarf planet.

Major Events in 2021

Jan. 3–4	Quadrantid meteor shower maximum
Jan. 21	Minor planet (15) Eunomia at opposition
Jan. 24	Mercury at greatest eastern elongation
Feb. 08	α-Centaurid meteor shower maximum
Mar. 04	Minor planet (4) Vesta at opposition
Mar. 06	Mercury at greatest western elongation
Mar. 14–15	γ-Normid meteor shower maximum
Apr. 17	Mars occulted by the Moon
Apr. 22	April Lyrid meteor shower maximum
Apr. 23–24	π-Puppid meteor shower maximum
May 06	η-Aquariid meteor shower maximum
May 17	Mercury at greatest eastern elongation
May 26	Total lunar eclipse
Jun. 10	Annular solar eclipse
Jul. 04	Mercury at greatest western elongation
Jul. 17	Minor planet (6) Hebe at opposition
Jul. 28	Piscis Austrinid meteor shower maximum
Jul. 30	α-Capricornid meteor shower maximum
Jul. 30	Southern δ-Aquariid meteor shower maximum
Aug. 12–13	Perseid meteor shower maximum
Aug. 20	Jupiter at opposition

Major events in 2021 (continued)

Sep. 01	α-Aurigid meteor shower maximum
Sep. 02	Saturn at opposition
Sep. 11	Minor planet (2) Pallas at opposition
Sep. 14	Mercury at greatest eastern elongation
Sep. 14	Neptune at opposition
Oct. 08–09	Draconid meteor shower maximum
Oct. 10	Southern Taurid meteor shower maximum
Oct. 21	Orionid meteor shower maximum
Oct. 25	Mercury at greatest western elongation
Oct. 29	Venus at greatest eastern elongation
Nov. 04	Uranus at opposition
Nov. 12	Northern Taurid meteor shower maximum
Nov. 17–18	Leonid meteor shower maximum
Nov. 19	Partial lunar eclipse
Nov. 27	Dwarf planet (1) Ceres at opposition
Dec. 02	Phoenicid meteor shower maximum
Dec. 04	Total solar eclipse
Dec. 07	Puppid-Velid meteor shower maximum
Dec. 14	Geminid meteor shower maximum
Dec. 22–23	Ursid meteor shower maximum

The Moon

The Moon at First Quarter.

The Moon

The monthly pages include diagrams showing the phase of the Moon for every day of the month, and also indicate the day in the *lunation* (or *age* of the Moon), which begins at New Moon. The diagrams showing the Moon's phase are repeated for southern-hemisphere observers who will see the Moon, south up. Although the main features of the surface – the light highlands and the dark maria (seas) – may be seen with the naked eye, far more features may be detected with the use of binoculars or any telescope. The many craters are best seen when they are close to the *terminator* (the boundary between the illuminated and the non-illuminated areas of the surface), when the Sun rises or sets over any particular region of the Moon and the crater walls or central peaks cast strong shadows. Most features become difficult to see at Full Moon, although this is the best time to see the bright ray systems surrounding certain craters. Accompanying the Moon map on the following pages is a list of prominent features, including the days in the lunation when features are normally close to the terminator and thus easiest to see. A few bright features such as Linné and Proclus, visible when well illuminated, are also listed. One feature, Rupes Recta (the Straight Wall) is readily visible only when it casts a shadow with light from the east, appearing as a light line when illuminated from the opposite direction.

The dates of visibility vary slightly through the effects of *libration*. Because the Moon's orbit is inclined to the Earth's equator and also because it moves in an ellipse, the Moon appears to rock slightly from side to side (and nod up and down). Features near the *limb* (the edge of the Moon) may vary considerably in their location and visibility. (This is easily noticeable with Mare Crisium and the craters Tycho and Plato.) Another effect is that at crescent phases before and after New Moon, the normally non-illuminated portion of the Moon receives a certain amount of light, reflected from the Earth. This *Earthshine* may enable certain bright features (such as Aristarchus, Kepler and Copernicus) to be detected even though they are not illuminated by sunlight.

Moon features

The numbers on the following page indicate the age of the Moon when features are usually best visible.

Abulfeda	6:20	Geminus	3:17	Pitatus	8:22
Agrippa	7:21	Goclenius	4:18	Pitiscus	5:19
Albategnius	7:21	Grimaldi	13–14:27–28	Plato	8:22
Aliacensis	7:21	Gutenberg	5:19	Plinius	6:20
Alphonsus	8:22	Hercules	5:19	Posidonius	5:19
Anaxagoras	9:23	Herodotus	11:25	Proclus	14:18
Anaximenes	11:25	Hipparchus	7:21	Ptolemaeus	8:22
Archimedes	8:22	Hommel	5:19	Purbach	8:22
Aristarchus	11:25	Humboldt	3:15	Pythagoras	12:26
Aristillus	7:21	Janssen	4:18	Rabbi Levi	6:20
Aristoteles	6:20	Julius Caesar	6:20	Reinhold	9:23
Arzachel	8:22	Kepler	10:24	Rima Ariadaeus	6:20
Atlas	4:18	Landsberg	10:24	Rupes Recta	8
Autolycus	7:21	Langrenus	3:17	Saussure	8:22
Barrow	7:21	Letronne	11:25	Scheiner	10:24
Billy	12:26	Linné	6	Schickard	12:26
Birt	8:22	Longomontanus	9:23	Sinus Iridum	10:24
Blancanus	9:23	Macrobius	4:18	Snellius	3:17
Bullialdus	9:23	Mädler	5:19	Stöfler	7:21
Bürg	5:19	Maginus	8:22	Taruntius	4:18
Campanus	10:24	Manilius	7:21	Thebit	8:22
Cassini	7:21	Mare Crisium	2–3:16–17	Theophilus	5:19
Catharina	6:20	Maurolycus	6:20	Timocharis	8:22
Clavius	9:23	Mercator	10:24	Triesnecker	6–7:21
Cleomedes	3:17	Metius	4:18	Tycho	8:22
Copernicus	9:23	Meton	6:20	Vallis Alpes	7:21
Cyrillus	6:20	Mons Pico	8:22	Vallis Schröteri	11:25
Delambre	6:20	Mons Piton	8:22	Vlacq	5:19
Deslandres	8:22	Mons Rümker	12:26	Walther	7:21
Endymion	3:17	Montes Alpes	6–8:21	Wargentin	12:27
Eratosthenes	8:22	Montes Apenninus	8	Werner	7:21
Eudoxus	6:20	Orontius	8:22	Wilhelm	9:23
Fra Mauro	9:23	Pallas	8:22	Zagut	6:20
Fracastorius	5:19	Petavius	3:17		
Franklin	4:18	Philolaus	9:23		
Gassendi	11:25	Piccolomini	5:19		

Map of the Moon

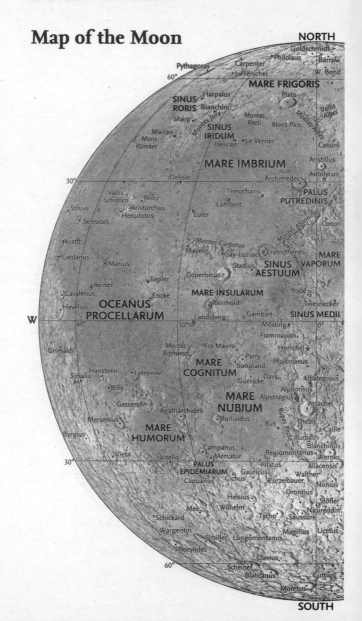

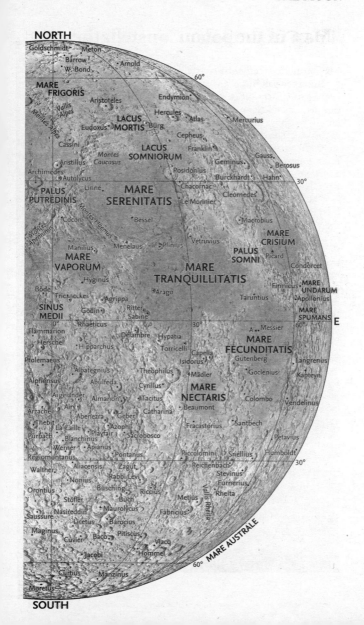

NORTH

Goldschmidt · Meton
Barrow · Arnold
W. Bond

MARE
FRIGORIS
Aristoteles
Vallis
Alpes
LACUS
MORTIS
Eudoxus
Cassini
Montes
Caucasus
LACUS
SOMNIORUM
Aristillus
Autolycus
Archimedes
Linné
PALUS
PUTREDINIS
Cocon
MARE
SERENITATIS
Bessel
Montes
Apenninus
Manilius
Menelaus
Plinius
MARE
VAPORUM
Hyginus
Bode
Triesnecker
Agrippa
SINUS
MEDII
Godin
Ritter
Sabine
Rhaeticus
Flammarion
Delambre
Hypatia
Herschel
Hipparchus
Torricelli
Ptolemaeus
Albategnius
Theophilus
Alphonsus
Abulfeda
Cyrillus
Argelander
Almanon
Tacitus
Arzachel
Airy
Abenezra
Geber
Thebit
La Caille
Azophi
Purbach
Blanchinus
Playfair
Werner
Apianus
Sacrobosco
Regiomontanus
Pontanus
Walther
Aliacensis
Zagut
Nonius
Rabbi Levi
Orontius
Busching
Stofler
Buch
Riccius
Nasireddin
Maurolycus
Saussure
Licetus
Barocius
Maginus
Cuvier
Baco
Pitiscus
Jacobi
Curtius
Manzinus
Moretus

SOUTH

Endymion
Hercules
Atlas
Mercurius
Burg
Cepheus
Franklin
Gauss
Geminus
Berosus
Posidonius
Burckhardt
Hahn
Chacornac
Cleomedes
Le Monnier
Macrobius
MARE
CRISIUM
Vetruvius
PALUS
SOMNI
Picard
Condorcet
MARE
TRANQUILLITATIS
Firmicus
MARE
UNDARUM
Arago
Taruntius
Apollonius
MARE
SPUMANS
Messier
MARE
FECUNDITATIS
Isidorus
Capella
Gutenberg
Langrenus
Madler
Goclenius
Kapteyn
MARE
NECTARIS
Colombo
Catharina
Beaumont
Vendelinus
Fracastorius
Santbech
Piccolomini
Snellius
Petavius
Humboldt
Reichenbach
Stevinus
Furnerius
Metius
Rheita
Vallis
Rheita
Fabricius
Vlacq
Hommel
MARE AUSTRALE

60°
30°
0°
30°
60°

E

25

The Circumpolar Constellations

The northern circumpolar constellations

Learning the patterns of the stars, the constellations and asterisms is not particularly difficult. You need to start by identifying the various constellations that are circumpolar where you live. These are always above the horizon, so you can generally start at any time of the year. The charts on pages 27 and 29 show the northern and southern circumpolar constellations, respectively. The fine, dashed lines indicate the areas that are circumpolar at different latitudes.

The key constellation when learning the pattern of stars in the northern sky is **Ursa Major**, in particular the seven stars forming the asterism, known to more-or-less everyone, as 'the **Plough**' or to people in North America, as the '**Big Dipper**'. As the chart shows, this is just circumpolar for anyone at latitude 40°N, except for **Alkaid** (η Ursae Majoris), the last star in the 'tail'. Even so, the asterism of the Plough is low on the northern horizon between September and November, so it will be much easier to make out at other times of the year.

A single day on **Venus** lasts 243.025 days. This rotation is longer than its orbital period, which is 224.701 days. Its rotation is retrograde: in the opposite direction to that of all the other planets.

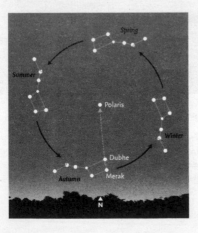

The two stars **Dubhe** and **Merak** (α and β Ursae Majoris) are known as the 'Pointers', because they indicate the position of **Polaris**, the Pole Star (α Ursae Minoris), at about a distance of five times their separation. Following this line takes you to the constellation of **Ursa Minor**, the 'Little Bear' or 'Little Dipper', where Polaris is at the end of the 'tail' or 'handle'. On the far side of the Pole is the constellation of **Cassiopeia**, which is

The northern circumpolar constellations.

highly distinctive, with its five main stars forming the letter 'M' or 'W', depending on its orientation. Cassiopeia is circumpolar for observers at latitude 40°N or closer to the North Pole, although at times it is near the northern horizon and more difficult to see. (But at such times Ursa Major is clearly visible.) To find Cassiopeia from Ursa Major, start at **Alioth** (ε Ursae Majoris) and extend a line from that star to Polaris and beyond. It points to the central star of the five.

Moving anticlockwise from Cassiopeia, we come to *Cepheus*, which has been likened to the gable end of a house, with its base in the Milky Way. The line from the Pointers to Polaris, if extended points to *Errai* (γ Cephei), at the top of the 'gable'. Continuing in the same direction, we come to *Draco,* which wraps around Ursa Minor. The quadrilateral of stars that forms the 'head' of Draco is just circumpolar for observers at 40°N latitude, although it is brushing the horizon in January. On the opposite side of the sky to the head of Draco, the whole of the faint constellation of *Camelopardalis* is visible.

For observers slightly farther north, say at 50°N, additional constellations become circumpolar. The most important of these are *Perseus*, not far from Cassiopeia, and most of which is visible and, farther round, the northern portion of *Auriga*, with bright *Capella* (α Aurigae). On the other side of the sky is *Deneb*, the brightest star in *Cygnus*, although it is often close to the horizon, especially during the early night during the winter months. *Vega* (α Lyrae) another of the three stars that form the Summer Triangle is even farther south, often brushing the northern horizon, and only truly circumpolar and clearly seen at any time of the year for observers at 60°N.

Such far northern observers will also find that *Castor* (α Geminorum) is actually circumpolar, although at times it is extremely low on the horizon. The other bright star in *Gemini*, *Pollux* (β Geminorum) is slightly farther south and cannot really be considered circumpolar.

The southern circumpolar constellations

Just as Ursa Major is the key constellation in the northern sky, so is *Crux* (the Southern Cross) an easily recognized feature of the southern circumpolar sky, although at times it may be brushing the horizon for observers at 30°S

The southern circumpolar constellations.

– roughly the latitude of Sydney in Australia. This is particularly true in the southern spring. Northeners, new to the southern sky, sometimes mistake the '**False Cross**', which consists of two stars from each of the constellations of **Vela** and **Carina** for the true Southern Cross. Crux itself is accompanied by the clearly visible dark cloud of the 'Coalsack' and also the two brightest stars in Centaurus: **Rigil**

Kentaurus and *Hadar* (α and β Centauri, respectively). Together, the four stars of Crux and the two from Centaurus act as principal guides to the southern constellations.

Unfortunately, unlike the situation in the north, there is no star conveniently located at the South Celestial Pole (SCP), which lies in a relatively empty region of sky in the faint, triangular constellation of *Octans*. Octans itself is perhaps best found by using the stars of *Pavo* as guides. A line from *Peacock* (α Pavonis) through the second brightest star (β) in that constellation, if extended by about the same amount as the distance between the two stars, points to the 'base' of the triangle of Octans.

The main 'upright' of Crux, if extended and curving slightly to the right, does point in the approximate direction of the Pole, passing through *Musca* and the tip of *Chamaeleon*. However, a better way is to start at Hadar (the star in the bright pair that is closest to Crux), turn at right-angles at Rigil Kentaurus, and following an imaginary line through the brightest star in the small constellation of

Olympus Mons, on Mars, is the highest mountain (volcano) in the Solar System. It is about 27 km high (measured as 21,287.4 metres above a datum level) and is roughly the size of Spain.

Circinus and then right across the sky, brushing past the outlying star of *Apus*, and the star (δ) at the apex of Octans itself. Centaurus is a straggling constellation, with many stars well north of Rigil Kentaurus and Hadar, and some fainter ones that partially enclose Crux. Starting at Crux, and moving clockwise (in the same direction as the sky rotates), we come to the stars of *Carina* and *Vela*, both originally part of the large, now obsolete constellation of Argo Navis. Past the False Cross, we come to *Canopus* (α Carinae), the second brightest star in the sky, which is just circumpolar for observers at 40°S, although occasionally, especially in June, very low over the northern horizon. Lying between Canopus and the SCP is the *Large Magellanic Cloud* (LMC), a satellite galaxy to our own. It lies across the boundaries of the constellations of *Dorado* and *Mensa*.

Continuing round from Canopus we pass the constellations of Dorado, the small constellation of *Reticulum* and the undistinguished constellation of *Horologium*, beyond which is *Achernar* (α Eridani) the brightest star in the long, winding constellation of *Eridanus*, which actually starts far to the north, close to *Rigel* in *Orion*. Between Achernar and the SCP, lies the triangular constellation of *Hydrus*, next to the constellation of *Tucana* which contains the *Small Magellanic Cloud* (SMC).

For observers farther south (at say, 50°S) there are the constellations of *Phoenix*, followed by the roughly cross-shaped constellation of *Grus* and the rather faint *Indus*. For observers at 40°S, the whole of the constellation of Pavo is visible, including its brightest star, Peacock. Farther round there is the constellation of *Ara* and, for observers farther north at 30°S, *Triangulum Australe* is fully visible.

The Monthly Maps

How to use the monthly maps

The charts in this book are designed to be used more-or-less anywhere in the world. They are not suitable to be used at very high northern or southern latitudes (beyond 60°N or 60°S). That is slightly less than the latitudes of the Arctic and Antarctic Circles, beyond which there are approximately six months of daylight, followed by six months of darkness. The design may seem a little complicated, but these diagrams should make their usage clear. The main charts are given in pairs, one pair for each month: Looking North and Looking South.

Obviously, the region of the sky that is visible at any time entirely depends on one's location on Earth. You should imagine a rectangular 'window', 90° degrees high, that includes the sky from the horizon to the zenith. Think of moving this 'window' north or south over the charts, depending on your actual latitude. The base will be at your actual latitude on Earth. The other edge will be at your zenith. (You could make an actual 'window', by drawing a rectangle on a thin sheet of plastic, with the horizon and zenith lines 90° apart, and then use this on the charts.) The diagram on the next page shows this 'window', for the latitude of 50°N.

The scales on the right-hand and left-hand margins indicate the northern (or southern) horizon, for looking North (or south) respectively. The two diagrams on page 34 are drawn to indicate the horizon for the latitude of 40°N (the latitude of Philadelphia in the United States or Madrid in Spain); the second pair on page 35, show what would be visible 'Looking North' and 'Looking South' from latitude 30°S (the latitude of Durban in South Africa). If you are looking north (or south), once you get to the zenith, you can switch to the other chart, showing the view from the southern (or northern) horizon to the zenith.

To help you choose the correct latitude, there is a world map on pages 36 and 37.

Horizon window, from the northern horizon to the zenith, for the latitude of 50°N.

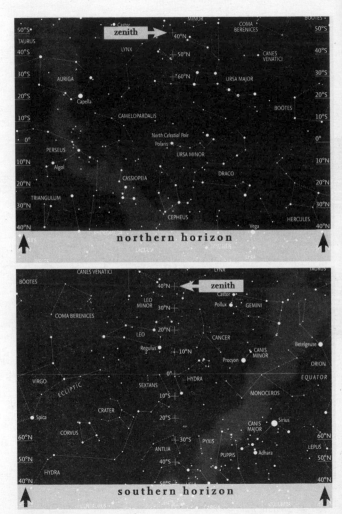

Horizons for latitude 40°N.

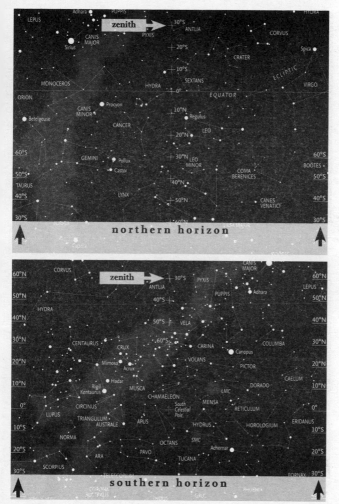

Horizons for latitude 30°S.

World Map

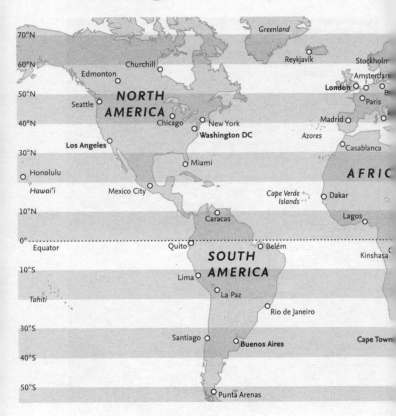

January

January – Introduction

The Earth
The Earth reaches perihelion (the closest point to the Sun in its annual orbit) on 2 January 2021, at 13:51 Universal Time. Its distance is then 0.983257 AU (147,093,163 km).

There are no major astronomical events during January 2021, but it is the month in which the peak of the annual Quadrantid meteor shower occurs (see page 52). This is a shower that is visible from the northern hemisphere only, with an extremely sharp peak in January. Unfortunately, because of the cold conditions and weather at that time of year, the shower tends to be poorly observed.

Occultations
Of the five bright stars near the ecliptic that may be occulted by the Moon (*Aldebaran*, *Antares*, *Pollux*, *Regulus*, and *Spica*), none are occulted in 2021. There is one occultation of *Mars*, on April 17, partly visible from SE Asia. There are occultations of the minor planets *Flora* (on November 30) and *Pallas* (on December 10), but both objects are very faint, mags. 11.5 and 9.9, respectively, so, although the existence of the events are noted in the relevant months, explicit details are not given.

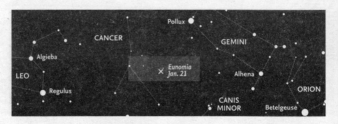

A finder chart for the position of minor planet (15) Eunomia, at its opposition. The grey area is shown in more detail on the facing page.

The Planets

In the evening sky, **Mercury** reaches greatest eastern elongation at mag. -0.7 on January 24. **Venus** (mag. -3.9), is initially in **Ophiuchus**, but soon moves into Sagittarius and closer to the Sun. **Mars** moves from **Pisces** to **Aries**, fading from mag -0.2 to 0.4. Mars does not come to opposition in 2021, and early January is the best time this year for viewing the planet. **Jupiter** (mag. -2.0) is in **Capricornus**, too close to the Sun to be visible. **Saturn**, also in Capricornus, is similarly invisible near the Sun. **Uranus** (mag. 5.7 to 5.8) is initially retrograding (moving westwards) in Aries, but resumes direct motion (moving east) on January 16. **Neptune** (mag. 7.9) is in **Aquarius**, where it remains for the whole year. On January 21, minor planet (15) **Eunomia** is at opposition (mag. 8.5) in **Cancer** (see the charts on page 40 and below).

On 14 January 2005, ESA's **Huygens** probe landed on the surface of Titan, having been carried to Saturn's system by the **Cassini** spaceprobe, from which it was released on 25 December 2004.

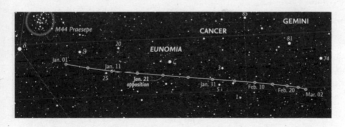

The path of the minor planet (15) Eunomia around its opposition on January 21 (mag. 8.5). Background stars are shown down to magnitude 9.5.

Sunrise and Sunset

City	Date	Sunrise	Sunset
Buenos Aires, Argentina			
	Jan. 01	08:44	23:10
	Jan. 31	09:13	23:00
Cape Town, South Africa			
	Jan. 01	03:39	18:01
	Jan. 31	04:07	17:52
London, UK			
	Jan. 01	08:07	16:02
	Jan. 31	07:41	16:49
Los Angeles, USA			
	Jan. 01	14:59	00:54
	Jan. 31	14:51	01:22
Nairobi, Kenya			
	Jan. 01	03:30	15:42
	Jan. 31	03:41	15:51
Sydney, Australia			
	Jan. 01	18:48	09:09
	Jan. 31	19:17	09:01
Tokyo, Japan			
	Jan. 01	21:51	07:38
	Jan. 31	21:41	08:07
Washington, DC, USA			
	Jan. 01	12:37	22:10
	Jan. 31	12:25	22:41
Wellington, New Zealand			
	Jan. 01	16:52	07:57
	Jan. 31	17:26	07:43

NB: the times given are in Universal Time (UT)

The Moon's Phases and Ages

Northern hemisphere

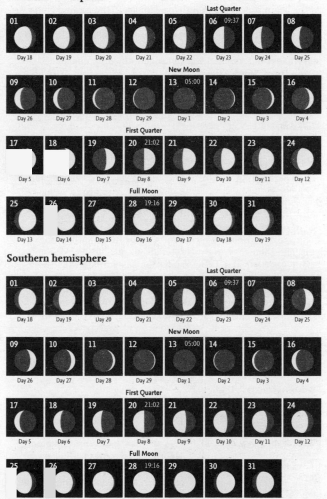

					Last Quarter		
01	02	03	04	05	06 09:37	07	08
Day 18	Day 19	Day 20	Day 21	Day 22	Day 23	Day 24	Day 25

				New Moon			
09	10	11	12	13 05:00	14	15	16
Day 26	Day 27	Day 28	Day 29	Day 1	Day 2	Day 3	Day 4

			First Quarter				
17	18	19	20 21:02	21	22	23	24
Day 5	Day 6	Day 7	Day 8	Day 9	Day 10	Day 11	Day 12

			Full Moon			
25	26	27	28 19:16	29	30	31
Day 13	Day 14	Day 15	Day 16	Day 17	Day 18	Day 19

Southern hemisphere

					Last Quarter		
01	02	03	04	05	06 09:37	07	08
Day 18	Day 19	Day 20	Day 21	Day 22	Day 23	Day 24	Day 25

				New Moon			
09	10	11	12	13 05:00	14	15	16
Day 26	Day 27	Day 28	Day 29	Day 1	Day 2	Day 3	Day 4

			First Quarter				
17	18	19	20 21:02	21	22	23	24
Day 5	Day 6	Day 7	Day 8	Day 9	Day 10	Day 11	Day 12

			Full Moon			
25	26	27	28 19:16	29	30	31
Day 13	Day 14	Day 15	Day 16	Day 17	Day 18	Day 19

The Moon

The Moon in January

At the beginning of the month (on January 2) the waning gibbous Moon passes 4.7° north of **Regulus**, the brightest star in **Leo**. On January 6, at Last Quarter, it is north of **Spica** in **Virgo**. On January 10, just before New Moon, it passes 6.0° north of the red supergiant star **Antares** in **Scorpius**. Later in the month, (on January 21) it is 5.1° south of **Mars**, which has moved from **Pisces** into **Aries**. Shortly afterwards, the Moon is 3.3° south of **Uranus**, also in **Aries**. On January 24 (when waxing gibbous) it passes 4.8° north of orange-tinted **Aldebaran** in **Taurus**, and then, on January 27 (one day before Full) it is south of **Pollux** in **Gemini**. On January 30, waning gibbous again, it is once more north of Regulus.

On 3 January 2019, the Chinese **Chang'e 4** spaceprobe became the first object to land on the far side of the Moon in the South Pole-Aitken Basin. It deployed the **Yutu-2** rover.

Wolf Moon

The Full Moon in January is sometimes known as the 'Wolf Moon', named, of course, after the howling of wolves, which may often be heard at that time of year. The name may originally stem from the Old-World, Anglo-Saxon lunar calendar. Other names for this Full Moon include: Moon After Yule, Old Moon, Ice Moon, and Snow Moon. Among the Algonquin tribes, the name was 'squochee kesos', meaning 'the Sun has not strength to thaw'. The name 'Wolf Moon' was occasionally applied to the Full Moon in December.

In this photograph, the narrow lunar crescent (about two days old) has been over-exposed to show the Earthshine illuminating the other portion of the Moon, where the dark maria are faintly visible.

Names of the Full Moon
The Native American tribes had a set of names for the Full Moon, depending on the time of year. The actual names varied between different tribes, so there may be more than one name used for a particular Full Moon. The interval between successive Full Moons (or between any other specific phases of the Moon) is known as the synodic month, and is, on average, 29.53 days, so the names have come to be associated with modern calendar months. They are:

- January: 'Wolf Moon'
- February: 'Snow Moon', also 'Hunger Moon'
- March: 'Worm Moon', 'Crow Moon', 'Sap Moon', 'Lenten Moon'
- April: 'Seed Moon', 'Pink Moon', 'Sprouting Grass Moon', 'Egg Moon', 'Fish Moon'
- May: 'Milk Moon', 'Flower Moon', 'Corn Planting Moon'
- June: 'Mead Moon', 'Strawberry Moon', 'Rose Moon', 'Thunder Moon'
- July: 'Hay Moon', 'Buck Moon', 'Elk Moon', 'Thunder Moon'
- August: 'Corn Moon', 'Sturgeon Moon', 'Red Moon', 'Green Corn Moon', 'Grain Moon'
- September: 'Harvest Moon', 'Full Corn Moon'
- October: 'Hunter's Moon', 'Blood Moon'/'Sanguine Moon'
- November: 'Beaver Moon', 'Frosty Moon'
- December: 'Oak Moon', 'Cold Moon', 'Long Night's Moon'

The only two names commonly used in Europe were 'Harvest Moon' and 'Hunter's Moon'. On rare occasions, particularly in religious contexts, the term 'Lenten Moon' was used for the Full Moon in March. The other terms, originating in North America, have been adopted increasingly by the media in recent years.

On 8 January 1973, the Soviet Union's **Lunokhod 2** rover landed in Le Monnier crater on the Moon. The mission ended when the signal was lost on 11 May 1973.

Calendar for January

01–12		Quadrantid meteor shower
02	13:51	Earth at perihelion (0.983257 AU = 147,093,163 km)
02	22:28	Regulus 4.7°S of Moon
03–04		Quadrantid shower maximum
06	09:37	Last Quarter
06	18:33	Spica 7.0°S of Moon
09	15:37	Moon at perigee (367,387 km)
09	21:00 *	Mercury 1.7°S of Saturn
10	02:38	Antares 6.0°S of Moon
11	11:00 *	Mercury 1.5°S of Jupiter
11	20:09	Venus 1.5°N of Moon
13	05:00	New Moon
13	20:52	Saturn 3.2°N of Moon
14	01:27	Jupiter 3.3°N of Moon
14	08:13	Mercury 2.3°N of Moon
17	06:14	Neptune 4.5°N of Moon
20	21:02	First Quarter
21	05:37	Mars 5.1°N of Moon
21	06:24	Uranus 3.3°N of Moon
21	13:11	Moon at apogee (404,360 km)
21	18:37	Minor planet (15) Eunomia at opposition (mag. 8.5)
22	00:00 *	Uranus 1.7°S of Mars
24	01:57	Mercury at greatest elongation (18.6°E, mag. -0.7)
24	03:01	Saturn in conjunction with the Sun
24	05:13	Aldebaran 4.8°S of Moon
27	16:18	Pollux 3.8°N of Moon
28	19:16	Full Moon
29	01:40	Jupiter in conjunction with the Sun
30	05:25	Regulus 4.6°S of Moon
31–Feb.20		α-Centaurid meteor shower

These objects are close together for an extended period around this time.

January 2 • *The Moon is halfway between Regulus and Algieba (γ Leo), as seen from London.*

January 9–11 • *Shortly before sunrise, the Moon passes Antares, Sabik (η Oph) and Venus (as seen from central USA).*

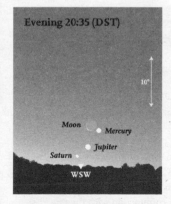

January 14 • *The narrow crescent Moon with Mercury, Jupiter and Saturn just after sunset (as seen from Sydney).*

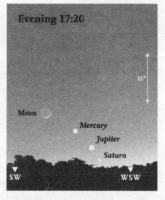

January 14 • *The narrow crescent Moon with Mercury, Jupiter and Saturn just after sunset (as seen from central USA).*

January – Looking North

For most northern observers, all the important northern circumpolar constellations (see pages 26 to 28) of **Ursa Major**, **Ursa Minor** (with **Polaris**, the Pole Star), **Cassiopeia**, **Draco** and **Cepheus** will be visible. Polaris is, of course, the star about which the sky appears to rotate, even though it is not precisely at the North Celestial Pole. Ursa Major stands more-or-less vertically above the horizon in the northeast. Opposite it in the northwest of the sky is the 'W' of Cassiopeia. (Looking more like an 'M' at this time of the year.) For most observers **Capella** (α Aurigae) is high overhead, but only those at low latitudes will find it easy to see the quadrilateral of stars that marks the head of Draco, brilliant **Deneb** (α Cygni) or the even brighter **Vega** (α Lyrae), yet farther south. Deneb and **Eltanin** (γ Draconis), the brightest star in the 'Head' of Draco, are skimming the horizon for observers at 40°N.

On 4 January 2004, the NASA surface rover, **Spirit**, landed within Gusev crater on Mars.

For observers between about 30 and 50°N, the constellation of **Auriga** is near the zenith (and thus difficult to observe). This important constellation contains bright **Capella** (α Aurigae) and farther to its west lies the constellation of **Perseus**, with **Algol**, the famous variable star.

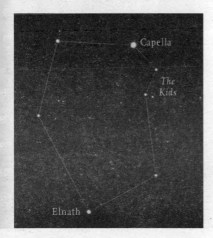

The constellation of Auriga, with brilliant Capella which, although appearing as a single star, is actually a quadruple system, consisting of a pair of yellow giant stars, gravitationally bound to a more distant pair of red dwarfs. Elnath, near the bottom edge, actually belongs to the constellation of Taurus.

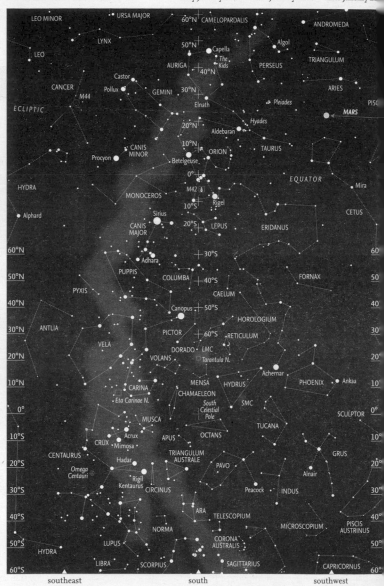

January – Looking South

The southern sky is dominated by *Orion*, visible from nearly everywhere in the world and prominent during the northern winter months. For observers near the equator it is, of course, high above near the zenith. Orion is highly distinctive, with a line of three stars that form the 'Belt'. To most observers, the bright star *Betelgeuse* (α Orionis), shows a reddish tinge, in contrast to the brilliant bluish-white *Rigel* (β Orionis). The three stars of the belt lie directly south of the celestial equator. A vertical line of three 'stars' forms the 'Sword' that hangs south of the Belt. With good viewing, the central 'star' appears as a hazy spot, even to the naked eye, and is actually the *Orion Nebula* (M42). Binoculars reveal the four stars of the Trapezium, which illuminate the nebula.

Orion's Belt points up to the northwest towards *Taurus* (the Bull) and orange-tinted *Aldebaran* (α Tauri). Close to Aldebaran, there is a conspicuous 'V' of stars, called the *Hyades* cluster. (Despite appearances, Aldebaran is not part of the cluster.) Farther along, the same line from Orion passes below a bright cluster of stars, the *Pleiades*, or Seven Sisters. Even the smallest pair of binoculars reveals this as a beautiful group of bluish-white stars. The two most conspicuous of the other stars in Taurus lie directly north of Orion, and form an elongated triangle with Aldebaran. The northernmost, *Elnath* (β Tauri), was once considered to be part of the constellation of *Auriga*.

On 24 January 1986 the *Voyager 2* spaceprobe had its encounter with Uranus, on its way to the outer Solar System.

Slightly to the west of Capella lies a small triangle of fainter stars, known as '*The Kids*'. (Ancient mythological representations of Auriga show him carrying two young goats.) Together with Elnath, the body of Auriga forms a large pentagon on the sky, with The Kids lying on the western side (see page 49).

Running south from Orion is the long constellation of *Eridanus* (the River), which begins near Rigel in Orion and runs far south to end at *Achernar* (α Eridani). To the south of Orion is the constellation of *Canis Major* and several other constellations, including the oddly shaped *Carina*. The line of Orion's Belt also points southeast in the general direction of *Sirius* (α Canis Majoris), the brightest star. Almost due south of Sirius lies *Canopus* (α Carinae), the second brightest star in the sky.

Meteors

Quadrantids

The Quadrantid meteor shower (named after the obsolete constellation of **Quadrans Muralis**) is active from the end of December (December 28) to January 12. It is one of the strongest showers of the year, with rates similar to those of the more prominent showers of the Perseids (in August) and Geminids (in December). There is, however, a very sharp peak, lasting no more than about 6 hours, and this next occurs on 3–4 January 2021. This sharp peak may be easily missed if there is bad weather or moonlight. (The Moon is waning gibbous in 2021, so conditions are not very favourable.) The overall brightness of the meteors is low, but it does frequently produce bright fireballs.

The radiant lies roughly half-way between θ Boötis and τ Herculis, and may also be envisaged as lying between **Alkaid** (η UMa) – the end of the 'tail' of **Ursa Major** (the end of the handle of the Big Dipper) – and the head of **Draco**. Because of the radiant's location, the shower is not well seen from the southern hemisphere, although occasional meteors may be seen from as far south as 50°S.

The parent body has been tentatively identified as the minor planet 2003 EH1, which may be linked to comet C/1490 Y1, observed by Far-Eastern astronomers (Chinese, Japanese, and Korean) some 530 years ago.

A brilliant, very late Quadrantid fireball, photographed by Denis Buczynski from Portmahomack, Ross-shire, Scotland, on 15 January 2018, at 23:44 UT.

Quadrans Muralis

The name of the Quadrantid meteor shower is the only remaining link with the obsolete constellation of **Quadrans Muralis** (the Mural Quadrant). The constellation was originally proposed, under the name '**Le Mural**', in the late-eighteenth century (1790s) by the French astronomer Joseph de Lalande, director of the observatory at the École Militaire in Paris. It was named after the mural quadrant, used in observatories at that period, which consisted of a quadrant, mounted on a north-south wall, and which was used for determining stellar positions.

The first depiction of the constellation was in Jean Fortin's *Atlas Céleste*, published in 1795. The name of the constellation remained in use until the late-nineteenth century. The Quadrantid meteor shower was first noticed by the Italian, Antonio Brucalassi, in 1825, and independently named after the constellation in 1839 by Adolphe Quetelet of Brussels Observatory in Belgium and Edward C. Herrick of Connecticut. The constellation was located above the head of Boötes and the area now forms part of that constellation.

On 25 January 2004, NASA's robotic rover **Opportunity** landed on Meridiani Planum on Mars.

Quadrans Muralis is shown under its original name of 'Le Mural' in this chart from Jean Fortin's Atlas Céleste.

The area of the former constellation of Quadrans Muralis, now part of the constellation of Boötes.

February

February – Introduction

February is a quiet month for astronomical events. For northern observers, there are no major meteor streams, but southern observers are treated to two showers. The most important of these are the Centaurids, which actually begin on January 31, and consist of two separate streams: the *α-Centaurids* and the *β-Centaurids*, with radiants lying near those two stars (Rigil Kentaurus and Hadar, respectively). Both showers reach a low maximum, with an hourly rate of about 5 meteors per hour, on February 8. That day the Moon is at Day 27 of the lunation, three days before New Moon, so observing conditions are very favourable.

The second shower, the *γ-Normids*, begins to be active around February 24 and continues into March, reaching its weak (but very sharp) maximum on March 14–15. Unfortunately, its meteors are difficult to differentiate from sporadics, so are likely to be identified by dedicated meteor observers only.

Did you know?
Why is February the shortest month? And why does it have such an odd number of days? The answers to these questions are surprisingly complicated. They involve the lunar calendar, ancient Romans, priests, Julius Caesar, and the way politicians tinkered with the calendar. A fairly comprehensive description of how these oddities came about is given on page 58–59.

The Planets

Mercury passes inferior conjunction on February 8. *Venus* is far too close to the Sun, and is thus invisible in the morning twilight. *Mars*, initially in *Aries*, is now fading rapidly (from mag. 0.4 to 0.9 over the month), and moves eastwards from Aries into *Taurus* at the very end of the month. *Jupiter* is in *Capricornus*, but remains too close to the Sun to be visible. *Saturn*, in the same constellation, is also lost in the morning twilight. *Uranus* (mag. 5.8) is in *Aries* and *Neptune* (mag. 7.9 to 8.0) remains in *Aquarius*.

The constellation of Orion dominates the sky during the first months of the year, and is a useful starting point for recognizing other constellations in the southern sky. Here, Betelgeuse, Rigel and the Orion Nebula are prominent.

February, the Shortest Month

Did you know?

Why is February the shortest month at 28 or 29 days? It all comes down to the fact that originally all calendars were lunar ones based on the Moon. The problem is that a fixed number of lunar months does not agree with a solar year. The time from Full Moon to Full Moon (or between any other pair of phases), known to astronomers as the synodic month, is slightly more than 29.5 solar days. The best fit for a single solar year is generally 365 days, although nowadays we occasionally add an extra day to keep the calendar in line with the Sun.

The trouble with February lies with the Romans. Originally, they used a lunar calendar, and the winter was considered to be without months. Then, about 713 BC, two months, January and February were added to the end of the year. (The year was supposed to begin at the spring equinox in March.) To keep the lunar calendar in step with the year, what are called intercalary months were sometimes inserted.

On 28 February 2007, the **New Horizons** spaceprobe received a gravity boost from Jupiter. The Galilean moons were imaged on 27–28 February.

However, the calendar was initially under the control of priests and, as might be expected, they fiddled things to their advantage. The calendar was used to determine when people could carry out business, and when ceremonies – very important in ancient Rome – could occur. The priests added intercalary months whenever they thought fit; for example, when more taxes were required. There was utter chaos and the calendar never agreed with the solar year.

Eventually, in what would now be called a 'coup', Julius Caesar became 'dictator for life'. He instigated a calendar reform, and his calendar subsequently became known as the Julian calendar. In this, months alternated between 30 and 31 days. This was slightly too long ($6 \times 30 + 6 \times 31 = 366$ days), so one day was removed from the last month of the year: February, giving it 29 days. An additional day was returned to February every four years (the leap year), to keep things in step with the Sun. One month, Quintilis, was named Julius, (our July) the month of Julius Caesar's birth.

But then the priests messed things up again. They started counting leap years as occurring every three years. So chaos ensued again. The error was corrected by the emperor Augustus and by AD 8 the matter had been solved and the months and the Sun

F

were in agreement. But then the Senate decided to rename one month in honour of Augustus – so the month of Sextilis became our August. Unfortunately, under Caesar's scheme that month had just 30 days, whereas that honouring Caesar himself (our July) had 31 days. Obviously Augustus had to have the same number of days, so they pinched one from poor February, leaving it with 28 days, except in leap years. At the same time, to avoid having three months with 31 days in succession (July, August and September) they also tinkered with the lengths of the months after August, which is why September and November now have 30 days and October and December 31.

The names of the months

Several of our modern names for the months derive from their order in the original Roman calendar. This, known as the 'calendar of Romulus' contained just ten months, because the winter was considered to be without Full Moons and thus months. This calendar dates back to around 750 BC. The legendary king was supposed to have started naming the months after gods, beginning at the spring equinox (in March). So we have:

- March from Martius (Mars, the god of war)
- April from Aprilis (the reason for this one uncertain, and possibly related to the raising of hogs)
- May from Maius (a local Italian goddess)
- June from Junius (the queen of the Latin gods).

After Junius, he simply started counting the months beginning

- Quintilis, for our July
- Sextilis for our August (See previous page for details of how the names of these two months changed.)

So we have:

- September from septem (seven)
- October from octo (eight)
- November from novem (nine)
- December from decem (ten)

Two, January (Januarius from Janus, the Roman god of beginnings) and February (Februarius, from Februa, the Roman ritual of purification at the end of the year) were later added at the end of the calendar.

Sunrise and Sunset

City	Date	Sunrise	Sunset
Buenos Aires, Argentina			
	Feb. 01	09:14	23:00
	Feb. 28	09:40	22:31
Cape Town, South Africa			
	Feb. 01	04:08	17:51
	Feb. 28	04:33	17:23
London, UK			
	Feb. 01	.07:39	16:50
	Feb. 28	06:48	17:40
Los Angeles, USA			
	Feb. 01	14:50	01:23
	Feb. 28	14:23	01:48
Nairobi, Kenya			
	Feb. 01	03:41	15:51
	Feb. 28	03:41	15:49
Sydney, Australia			
	Feb. 01	19:18	09:00
	Feb. 28	19:43	08:33
Tokyo, Japan			
	Feb. 01	21:40	08:08
	Feb. 28	21:11	08:35
Washington, DC, USA			
	Feb. 01	12:14	22:30
	Feb. 28	11:42	23:00
Wellington, New Zealand			
	Feb. 01	17:28	07:42
	Feb. 28	18:02	07:06

NB: the times given are in Universal Time (UT)

The Moon's Phases and Ages

Northern hemisphere

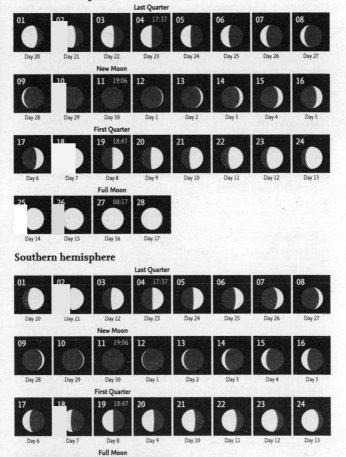

Last Quarter

01	02	03	04 17:37	05	06	07	08
Day 20	Day 21	Day 22	Day 23	Day 24	Day 25	Day 26	Day 27

New Moon

09	10	11 19:06	12	13	14	15	16
Day 28	Day 29	Day 30	Day 1	Day 2	Day 3	Day 4	Day 5

First Quarter

17	18	19 18:47	20	21	22	23	24
Day 6	Day 7	Day 8	Day 9	Day 10	Day 11	Day 12	Day 13

Full Moon

25	26	27 08:17	28
Day 14	Day 15	Day 16	Day 17

Southern hemisphere

Last Quarter

01	02	03	04 17:37	05	06	07	08
Day 20	Day 21	Day 22	Day 23	Day 24	Day 25	Day 26	Day 27

New Moon

09	10	11 19:06	12	13	14	15	16
Day 28	Day 29	Day 30	Day 1	Day 2	Day 3	Day 4	Day 5

First Quarter

17	18	19 18:47	20	21	22	23	24
Day 6	Day 7	Day 8	Day 9	Day 10	Day 11	Day 12	Day 13

Full Moon

25	26	27 08:17	28
Day 14	Day 15	Day 16	Day 17

F

The Moon

The Moon in February

On February 3, just after midnight one day before Last Quarter, the Moon passes 6.8° north of *Spica* (α Virginis) in the constellation of *Virgo*. On February 6, it is 5.4° north of red *Antares* in *Scorpius*. On February 18 it passes 3.7° south of *Mars* in *Aries*. Two days later, it passes south of the bright blue stars in the *Pleiades* cluster (M45) and then 5.0° north of orange-tinted *Aldebaran* in *Taurus*. On January 24 it is 3.7° south of *Pollux* in *Gemini*, and on February 26, 4.6° north of *Regulus* in *Leo*.

Snow Moon

In the northern hemisphere, February is often the coldest month, and most countries on both sides of the Atlantic see significant falls of snow. The Full Moon of February is thus often called the 'Snow Moon', although just occasionally that name has been applied to the Full Moon in January. Some North American tribes named it the 'Hunger Moon' because of the scarcity of food sources during the depths of winter, while other names are 'Storm Moon' and 'Chaste Moon', although the last name is more commonly applied to the Full Moon in March. To the Arapaho of the Great Plains, the Full Moon was called the Moon 'when snow blows like grain in the wind'.

Did you know?

Because it never has more than 29 days, and the synodic month (between any pair of phases, such as from New Moon to New Moon) is slightly more than half a day longer, sometimes there is no Full Moon in February. This occurs about every 19 years. This is one of the definitions of the term '*Black Moon*', although the same term is sometimes applied to a New Moon that does not occur in a particular season, as reckoned from the equinoxes or solstices.

F

Apollo 14

February 2021 sees the 50th anniversary of the Apollo 14 landing on the Moon. This mission, launched on 31 January 1971, landed on the Moon on 5 February, just north of the large crater of Fra Mauro (the intended target of the failed Apollo 13 mission) in the lunar highlands. A large amount of lunar rocks (over 42 kg) were collected and returned to Earth. One large sample ('Big Bertha') contains a meteorite, originating from Earth, that is the oldest known Earth rock, at about 4 billion years old.

The rock known as 'Big Bertha' in the laboratory on Earth. This, brought back by Apollo 14, is the third largest individual rock returned by the Apollo missions and weighs 8.998 kg.

The Apollo 14 landing site was outside the northern flank of the crater Fra Mauro, which is about 45 km across.

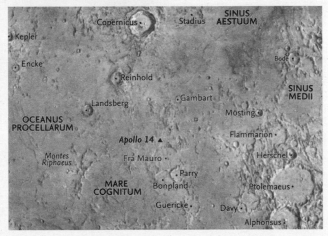

Calendar for February

03	00:02	Spica 6.8°S of Moon
03	19:03	Moon at perigee (370,127 km)
04	17:37	Last Quarter
06	05:00 *	Saturn 0.4°N of Venus
06	09:04	Antares 5.4°S of Moon
08		α-Centaurid shower maximum
08	13:48	Mercury at inferior conjunction
10	11:10	Saturn 3.4°N of Moon
10	20:25	Venus 3.2°N of Moon
10	21:35	Jupiter 3.7°N of Moon
11	03:17	Mercury 8.3°N of Moon
11	12:00 *	Jupiter 0.4°N of Venus
11	19:06	New Moon
12	17:00 *	Mercury 4.8°N of Venus
13	16:57	Neptune 4.3°N of Moon
13	19:00 *	Mercury 4.2°N of Jupiter
17	15:48	Uranus 3.0°N of Moon
18	10:22	Moon at apogee (404,467 km)
18	22:46	Mars 3.7°N of Moon
19	18:47	First Quarter
20	13:50	Aldebaran 5.0°S of Moon
24–Mar.24		γ-Normid meteor shower
24	01:42	Pollux 3.7°N of Moon
26	14:33	Regulus 4.6°S of Moon
27	08:17	Full Moon

These objects are close together for an extended period around this time.

February 6 • *In the early morning, the waning crescent Moon is five degrees north of Antares (as seen from London).*

February 18–20 • *The First Quarter Moon passes Mars, the Pleiades and Aldebaran in the west (as seen from London).*

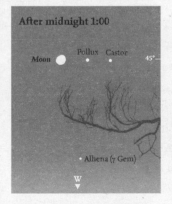

February 24 • *The Moon lines up with Castor and Pollux in the western sky, in the early morning (as seen from central USA).*

February 26 • *The Moon is almost full when it passes between Regulus and Algieba (γ Leo), as seen from London.*

northwest　　　　　　　north　　　　　　　northeast

February – Looking North

The months of January and February are probably the best time for seeing the section of the Milky Way that runs in the northern and western sky from **Orion** and **Gemini** right through **Auriga**, **Perseus** and **Cassiopeia**, towards **Cygnus**, low in the north. Although not as readily visible as the denser star clouds of the summer Milky Way, on a clear night so many stars may be seen that even a distinctive constellation such as Cassiopeia, which lies across the Milky Way, is not immediately obvious.

The 'base' of the constellation of **Cepheus** lies on the edge of the stars of the Milky Way, but the red supergiant star **Mu (μ) Cephei**, called the 'Garnet Star' by William Herschel – whose biography is on page 81 – with its striking red colour remains readily visible. The groups of stars, known as the **Double Cluster** in Perseus (NGC 869 & NGC 884, often known as h and χ Persei), lying between Perseus and Cassiopeia, are well-placed for observation.

Beyond the Milky Way, Perseus and Cassiopeia, the constellation of **Andromeda** is beginning to be lost in the north-western sky.

Castor and **Pollux** in **Gemini** are near the zenith for observers at 30°N, but for most northern observers the constellation is best seen when facing south. The head of **Draco** is now higher in the sky and easier to recognize, with its long, straggling 'body' curling round **Ursa Minor** and the Pole. **Ursa Major** has begun to swing round towards the north, and Cassiopeia is now lower in the western sky.

For observers in the far north, most of **Cygnus**, with its brightest star, **Deneb** (α Cygni), is visible in the north, and even **Lyra**, with **Vega** (α Lyrae) may be seen at times. Most of the constellation of **Hercules** is visible, together with the distinctive circlet of **Corona Borealis** to its west. Observers farther south may see Deneb and even Vega peeping over the northern horizon at times during the night, although they will often be lost (like all the fainter stars) in the inevitable extinction along the horizon.

On 3 February 1966, **Luna 9** became the first object to make a soft landing on the Moon. It returned data until 6 February.

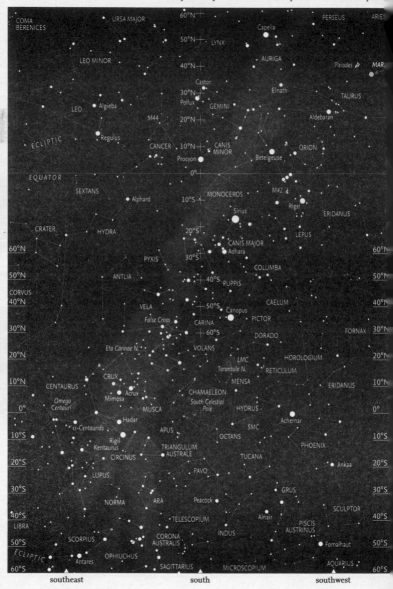

southeast   south   southwest

February – Looking South

Orion is now beginning to sink into the southwest, and the two brightest stars in the sky, **Sirius** and **Canopus** (α Carinae), are readily visible to observers at low northern latitudes and, of course, to those who are south of the equator. (Canopus is close to the zenith for those in the far south.)

South of **Carina** and the neighbouring constellation of **Vela** (both part of the original, and now obsolete, constellation of Argo Navis), lies the sprawling constellation of **Centaurus**, surrounding the distinctive constellation of **Crux**, the Southern Cross, which is on the horizon at 10°N. North of this, the 'False Cross', sometimes mistaken for the true constellation of Crux, consists of two stars from each of Vela (δ and κ Velorum) and Carina (ε and ι Carinae). The two brightest stars of Centaurus, **Rigil Kentaurus** (α Centauri) and **Hadar** (β Centauri) are slightly farther south, beyond Crux.

The **Large Magellanic Cloud** (LMC) is almost on the meridian, early in the night, to the west of Carina and the small constellation of **Volans**. It is surrounded by several small constellations: **Mensa**, **Reticulum** and **Dorado**. (Technically, it is largely within the area of Dorado.) Any optical aid (such as binoculars) will begin to show some of the remarkable structures within the LMC, including the great **Tarantula Nebula**, or **30 Doradus,** a site of active star formation.

Farther south and west lies **Achernar** (α Eridani), the bright star at the end of **Eridanus**, the long straggling constellation that represents a river and that may now be traced all the way from where it begins near **Rigel** (β Orionis) in Orion.

Beyond Crux, and on the other side of the Milky Way, lies the rest of Centaurus. Northeast of Crux is the finest and brightest globular cluster in the sky, **Omega** (ω) **Centauri** also known as NGC 5139. It is the largest globular cluster in our Galaxy and is estimated to contain about 10 million stars. Although appearing like a star, its non-stellar nature was discovered by Edmond Halley in 1677.

On 14 February 1972, the robotic probe **Luna 20** returned the first lunar samples, from the Apollonius highlands.

Mercury

The planet Mercury passes inferior conjunction with the Sun on February 8. In this case, the planet does not pass directly between the Earth and the Sun, in a *transit*. Transits of Mercury occur about 13 or 14 times every century, always in May or November (when the orbits are aligned) and the last two were in 2016 (May 9) and 2019 (November 11). The next one will be on 13 November 2032. Historically, transits of Mercury were important in attempts to determine the scale of the Solar System. They were also used to investigate the possible existence of the planet Vulcan, a hypothetical planet once thought to orbit the Sun inside the orbit of Mercury.

Mercury (the small black dot) in transit across the Sun on 8 November 2006. The sunspots are considerably larger than Mercury and the one on the left is larger than the diameter of the Earth.

Vulcan

The planet Vulcan was, for a period, believed to exist, with an orbit around the Sun inside the orbit of Mercury. The idea was originally raised by Urbain Le Verrier, a French mathematician (later Director of the Paris Observatory), who had previously calculated the position of the planet Neptune, which led to its discovery in 1846. He attempted to explain how the peculiar behaviour of the orbit of Mercury – now known to be a result of the effects of general relativity, as postulated by Albert Einstein – was caused by an additional planet.

Le Verrier developed a provisional theory for the motion of Mercury and this was tested at a transit of Mercury in 1848. Le Verrier's prediction failed to agree with the observations. He refined his calculations and, to account for the discrepancies in Mercury's orbit, postulated the existence of a small planet (Vulcan) moving within the orbit of Mercury. In 1859, a French doctor, Edmond Modeste Lescarbault, an amateur astronomer, wrote to Le Verrier, claiming to have seen a small planet crossing the Sun earlier in that year. Initially, he had thought the object to be a sunspot, but then believed it had moved, suggesting it was a tiny planet. He made hurried, crude measurements of its positions and timing.

Le Verrier accepted Lescarbault's observation, and announced the discovery of the planet, Vulcan. Le Verrier continued to believe in the existence of Vulcan, and repeatedly modified his theory to account for the various 'observations' of objects in the vicinity of the Sun, or crossing it in transits that were then reported, some by reputable astronomers. With the publication of Einstein's theory of relativity, which accounted for the peculiar behaviour of Mercury's orbit, and confirmation of that theory in the eclipse of 1919, most astronomers concluded that no significant planet could exist within the orbit of Mercury. Those observations that appeared to relate to a real object were thought to be associated with unknown minor planets or comets.

The suggestion has been made that there might be 'vulcanoids', tiny minor planets orbiting the Sun, but this has been largely discounted by intensive examination of the results from the **SOHO** (Solar and Heliospheric Observatory) spacecraft and the two **STEREO** (Solar Terrestrial Relations Observatory) spaceprobes. Although there may be very tiny objects orbiting the Sun, no vulcanoids larger than 5.7 km in diameter exist.

March

March – Introduction

On Saturday, March 20, the Sun crosses the celestial equator from south to north. The point at which this occurs, known as the *First Point of Aries*, was originally named many centuries ago when it lay in the constellation of *Aries*, and now (because of the phenomenon of precession, which changes the inclination of the Earth's axis of rotation) it actually lies in the constellation of *Pisces* and is close to the border with the constellation of Aquarius southeast of the star λ Piscium. It is moving towards *Aquarius* at the rate of about one degree every 70 years, and will eventually enter that constellation in some hundreds of years. (The *Age of Aquarius*, made popular by the song, is an astrological idea, rather than a true astronomical term and, astronomically, is rather premature.)

Day and night at the equinoxes are of almost equal length, and the northern hemisphere's season of spring is considered to begin in March. (Although, theoretically, the amount of time is the same, in fact refraction in the atmosphere 'raises' the position of the Sun at sunset and sunrise, so daylight lasts slightly longer than darkness.) The hours of daylight and night-time darkness change most rapidly around the equinoxes in March and September. The corresponding equinox in autumn (for the northern hemisphere, the spring equinox for those in the south), when the Sun crosses the celestial equator from north to south, occurs on September 22 in 2021, and always lies in the constellation of *Virgo*.

The First Point of Aries is sometimes known as the Cusp of Aries and the other equinox, in September, as the Cusp of Virgo.

The planet Uranus was discovered on 13 March 1781, by William Herschel, working from his house in Bath in Somerset. The discovery is described on page 77, and Herschel's career on page 81.

The Planets

On March 6, **Mercury** is at greatest western elongation 27.3° away from the Sun, low in morning twilight. **Venus** is also invisible in the twilight as it moves from the morning to the evening sky. It is at superior conjunction on the far side of the Sun on March 26. During the month, **Mars** fades slightly from mag. 0.9 to 1.3 as it moves eastwards across **Taurus**. **Jupiter** is lost in the morning twilight in **Capricornus** as is **Saturn**. **Uranus** (mag. 5.8) has just left **Pisces**, and is now in **Aries**, close to the Sun. **Neptune** remains in **Aquarius**, also close to the Sun, passing superior conjunction on March 11.

The minor planet *(4) Vesta* comes to opposition on March 4 in the constellation of **Leo**. Charts showing its location from the beginning of the year until the end of May, and around opposition are shown below, and Vesta is described in detail on page 76.

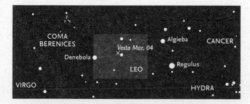

A finder chart for the position of minor planet (4) Vesta, at its opposition. The grey area is shown in more detail on the map below.

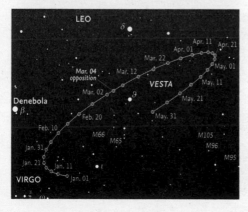

The path of the minor planet (4) Vesta around its opposition on March 4 (mag. 6.0). Background stars are shown down to magnitude 7.5.

M

Minor Planet Vesta

The minor planet *(4) Vesta* comes to opposition on 4 March 2021 at magnitude 0.6. Vesta is the second largest minor planet, after the dwarf planet *(1) Ceres*. It was discovered on 29 March 1807, by the German astronomer, Heinrich Olbers. It is believed to be slightly larger and more massive than the body known as *(2) Pallas*. It is one of the two minor planets visited by the Dawn spacecraft (the other being Ceres). Vesta is notable for being of a rocky nature and is suspected to be a primordial body, similar to those from which the inner, rocky planets (Mercury, Venus, Earth & Mars) originally formed. The body has undergone several major impacts, which have ejected fragments of rocky material into space. Some of these have fallen to Earth as meteorites, where they were long known to be of a specific kind (howardite-eucrite-diogenites) and suspected (but now confirmed) to come from Vesta.

A comparison of the three largest minor planets. 1 Ceres (left) is now termed a dwarf planet, partly because of its spherical shape, with a mean diameter of 939 km. 4 Vesta (centre) and 2 Pallas (right) are not spherical, having dimensions of about 572 × 557 × 446 km and 550 × 516 × 475 km, respectively. Ceres and Vesta were visited by the Dawn spacecraft.

Uranus

Uranus was discovered by William Herschel (later Sir William) on 13 March 1781, observing from his home at 19 New King Street in Bath, Somerset, UK. (This is now the Herschel Museum of Astronomy.) The planet is at the limit of naked-eye visibility and had been observed many times previously, but mistaken for a star. The earliest recorded observation may even be that by Hipparchos in 128 BC. It was observed by the famous astronomer John Flamsteed (in 1690) and no less than twelve times by the French astronomer Pierre Le Monnier (between 1750 and 1769). In his journal, Herschel recorded the observation as that of either a '... Nebulous star or perhaps a comet'. He tended to describe the object as a comet, although other astronomers suspected its planetary nature almost immediately. In 1781, the German astronomer, Johann Elert Bode, suggested it was a planet. He was also the first to suggest the name Uranus for the object, rather than Herschel's 'Georgium Sidus'. By 1783, Herschel himself had accepted that Uranus was a planet, making it the first planet to be discovered since antiquity. The first publication to definitely indicate its planetary nature was published in 1787 by the Finnish–Swedish astronomer, Anders Lexell, who calculated that it had a nearly circular orbit. Uranus comes to opposition on 4 November 2021 (see page 197).

The sizes of the Earth (left) and Uranus (right) compared.

Sunrise and sunset

City	Date	Sunrise	Sunset
Buenos Aires, Argentina			
	Mar. 01	09:41	22:30
	Mar. 31	10:05	21:49
Cape Town, South Africa			
	Mar. 01	04:34	17:22
	Mar. 31	04:58	16:42
London, UK			
	Mar. 01	06:46	17:41
	Mar. 31	05:38	18:33
Los Angeles, USA			
	Mar. 01	14:21	01:49
	Mar. 31	13:41	02:13
Nairobi, Kenya			
	Mar. 01	03:41	15:49
	Mar. 31	03:34	15:40
Sydney, Australia			
	Mar. 01	19:44	08:32
	Mar. 31	20:07	07:52
Tokyo, Japan			
	Mar. 01	21:10	08:36
	Mar. 31	20:28	09:02
Washington, DC, USA			
	Mar. 01	11:40	23:01
	Mar. 31	10:54	23:31
Wellington, New Zealand			
	Mar. 01	18:03	07:04
	Mar. 31	18:36	06:14

NB: the times given are in Universal Time (UT)

The Moon's phases and ages

Northern hemisphere

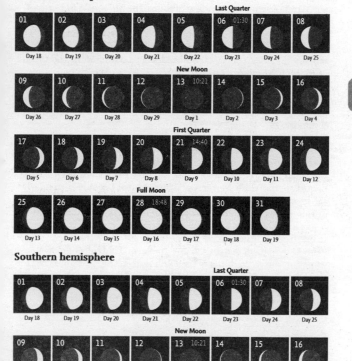

Last Quarter

					06 01:30		
01	02	03	04	05	06	07	08
Day 18	Day 19	Day 20	Day 21	Day 22	Day 23	Day 24	Day 25

New Moon

09	10	11	12	13 10:21	14	15	16
Day 26	Day 27	Day 28	Day 29	Day 1	Day 2	Day 3	Day 4

First Quarter

17	18	19	20	21 14:40	22	23	24
Day 5	Day 6	Day 7	Day 8	Day 9	Day 10	Day 11	Day 12

Full Moon

25	26	27	28 18:48	29	30	31
Day 13	Day 14	Day 15	Day 16	Day 17	Day 18	Day 19

M

Southern hemisphere

Last Quarter

01	02	03	04	05	06 01:30	07	08
Day 18	Day 19	Day 20	Day 21	Day 22	Day 23	Day 24	Day 25

New Moon

09	10	11	12	13 10:21	14	15	16
Day 26	Day 27	Day 28	Day 29	Day 1	Day 2	Day 3	Day 4

First Quarter

17	18	19	20	21 14:40	22	23	24
Day 5	Day 6	Day 7	Day 8	Day 9	Day 10	Day 11	Day 12

Full Moon

25	26	27	28 18:48	29	30	31
Day 13	Day 14	Day 15	Day 16	Day 17	Day 18	Day 19

The Moon

The Moon in March

On March 2, when it is waning gibbous, the Moon is 6.6° north of *Spica* in *Virgo*. On March 5, just before Last Quarter, the Moon is 5.2° north of the red supergiant *Antares* in *Scorpius*. A fortnight later, on March 19, as a waxing crescent on Day 7 of the lunation, the Moon passes south of the *Pleiades* cluster and *Mars*, and then north of *Aldebaran* in *Taurus*. On March 23, waxing gibbous, the Moon is 3.5° south of *Pollux* in *Gemini*. On March 26 (two days before Full Moon) it is 4.7° north of *Regulus* in *Leo*.

Worm Moon

One common name for the last Full Moon of the winter season, which falls in March, is the 'Worm Moon'. The name derives from the fact that earthworms become active in the soil at the end of winter and are sometimes seen at the surface of the soil. Other names are the 'Crow Moon', because the birds become particularly active and are avid to feed on the worms, after the lack of food during the winter months. Another name is the 'Sap Moon', which is particularly relevant in Canada, because this is the time when maple trees may be tapped for the sap (to produce the maple syrup, beloved by Canadians). In Europe, the term 'Lenten Moon' was sometimes used, and this is the Old English/Anglo-Saxon name for this particular Full Moon.

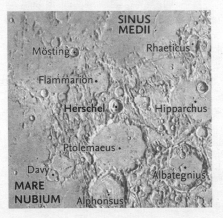

The crater, named after William Herschel (centre of the image) is about 40 km in diameter, with prominent inner terraces.

Sir William Herschel was born Friedrich Wilhelm Herschel in
Hanover on 15 November 1738. After a period as a musician in
the Hanoverian Military Band, Herschel emigrated to England
when he was nineteen. After initially acting as a musician in
Sunderland, Newcastle, Leeds and Halifax, he eventually moved
to Bath and became organist at the Octagon Chapel. He became
increasingly interested in astronomy, constructing his own
telescope, with which he discovered Uranus on 13 March 1781.
Initially believing the object to be a comet, by 1783 Herschel
accepted it as the first planet to be discovered since antiquity.
In 1782 he was appointed 'King's Astronomer' (or 'Court
Astronomer') by King George III. (This position must not be
confused with that of the Astronomer Royal, based at Greenwich
Observatory, who, at that time, was Nevil Maskelyne.) Herschel
moved to Slough, where he continued his astronomical work and
discoveries, ably assisted by his sister, Caroline, a considerable
astronomer in her own right. He made many other discoveries,
including the existence of infrared radiation. He became a Knight
of the Royal Guelphic Order, and also a Fellow of the Royal Society.
He was the first President of the Astronomical Society (later the
Royal Astronomical Society) on its formation in 1820. He died
in August 1822 and is buried in Slough. A crater on the Moon is
named after him as is minor planet (2000) Herschel.

M

*A faithful replica of the telescope with which
William Herschel discovered Uranus.*

Calendar for March

02	05:18	Moon at perigee (365,423 km)
02	06:58	Spica 6.6°S of Moon
04	18:08	Minor planet (4) Vesta at opposition (mag. 6.0)
05	07:00 *	Mercury 0.3°N of Jupiter
05	14:29	Antares 5.2°S of Moon
06	01:30	Last Quarter
06	11:22	Mercury at greatest elongation (27.3°W, mag. 0.1)
09	22:57	Saturn 3.7°N of Moon
10	15:36	Jupiter 4.1°N of Moon
11	00:01	Neptune in conjunction with the Sun
11	01:01	Mercury 3.7°N of Moon
11	12:00 *	Jupiter 0.4°N of Venus
13	00:16	Venus 3.9°N of Moon
13	02:38	Neptune 4.3°N of Moon
13	10:21	New Moon
14–15		γ-Normid shower maximum
17	01:50	Uranus 2.7°N of Moon
18	05:03	Moon at apogee (405,253 km)
19	17:47	Mars 1.9°N of Moon
19	21:48	Aldebaran 5.3°S of Moon
20	09:37	March equinox
21	14:40	First Quarter
23	21:00 *	Mars 7.0°N of Aldebaran
23	10:58	Pollux 3.5°N of Moon
26	00:46	Regulus 4.7°S of Moon
26	06:58	Venus at superior conjunction
28	18:48	Full Moon
29	16:22	Spica 6.5°S of Moon
30	06:16	Moon at perigee (360,309 km)

These objects are close together for an extended period around this time.

March 5–6 • The Moon passes between Antares and Sabik (η Oph), low in the south-southeast (as seen from London).

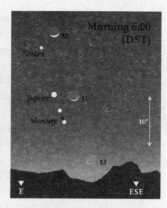

March 10–12 • The crescent Moon passes Saturn, Jupiter and Mercury in the east, shortly before sunrise (as seen from Sydney).

March 19 • The Moon is between Mars (mag. 1.1) and Aldebaran (mag. 0.9) in the western sky (as seen from London).

March 19 • The Moon is between Mars (mag. 1.1) and Aldebaran (mag. 0.9) in the western sky (as seen from central USA).

March – Looking North

The highly distinctive (and widely recognized) constellation of **Ursa Major** with the distinctive asterism of the Plough (or Big Dipper) is now 'upside down' and near the zenith for observers in the far north, for whom it is particularly difficult to observe. At this time of year, it is high in the sky for anyone north of the equator. Only observers farther towards the south will find it lower down towards their northern horizon and reasonably easy to see. However, at 30°S, even the seven stars making up the main, easily recognized portion of the constellation, are too low to be visible.

Auriga, with brilliant **Capella** (α Aurigae) is also very high on the opposite side of the meridian. The constellation of **Perseus** lies between it and **Andromeda** on the western side of the sky.

Ursa Minor, also with seven main stars, one of which is **Polaris**, the Pole Star, and the long constellation of **Draco** that winds around the Pole, are readily visible for anyone in the northern hemisphere, although, of course, Polaris is right on the horizon for anyone at the equator, and thus always lost to sight. **Cepheus** is near the meridian to the north, with **Cassiopeia**, to its west beginning to turn and resume its 'W' shape. The constellation of Andromeda is now diving down into the northwestern sky. In the east, beyond **Alkaid** (η Ursae Majoris), the final star in the 'tail' of Ursa Major, lies the top of **Boötes**. Farther to the south, most of **Hercules** and the 'Keystone' shape that forms the major portion of the body is visible.

Observers at 50°N may occasionally be able to detect bright **Deneb** (α Cygni) and **Vega** (α Lyrae) skimming the horizon, together with portions of those particular constellations, although most of the time they will be lost in the extinction that occurs at such low altitudes.

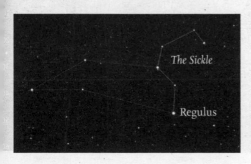

The distinctive constellation of Leo, with Regulus and 'The Sickle' on the west.

southeast south southwest

March – Looking South

The distinctive constellation of *Leo* (see page 85) is close to the meridian early in the night, clearly visible for anyone north of the equator. It is easily recognized, with bright *Regulus* (α Leonis) at the base of the 'backward question mark' of the asterism known as 'the Sickle', which lies north of Regulus. The minor planet *(4) Vesta* is at opposition in Leo on March 4 (see page 75). To the west of Leo is the inconspicuous constellation of *Cancer*, and still farther away from the meridian, the far more striking constellation of *Gemini*, with the bright stars *Castor* (α Geminorum), the closer to the Pole, and *Pollux* (β Geminorum). The constellation straddles the ecliptic, and Pollux may sometimes be occulted by the Moon (as may Regulus), although no such occultation occurs in 2021. Castor is remarkable in that even a fairly small telescope will show it as consisting of three stars (two fairly bright, and one fainter). However, more detailed investigation reveals that each of those stars is actually a double, so the whole system consists of no fewer than six stars.

Below Cancer is the very distinctive asterism of the 'Head of Hydra', consisting of five (or six) stars, that is the western end of the long constellation of *Hydra*, the largest constellation in the sky, that runs far towards the east, roughly parallel to the ecliptic. *Alphard* (α Hydrae) is south, and slightly to the west of Regulus in Leo and is relatively easy to recognize as it is the only fairly bright star in that region of the sky. North of Hydra and between it and the ecliptic and

Yuri Gagarin (19 March 1924 – 27 March 1968) was the first human to orbit the Earth, making a single orbit on 12 April 1961.

the constellation of *Virgo* are the two constellations of *Crater* and *Corvus*. Farther west, the small constellation of *Sextans* lies between Hydra and Leo.

Farther south, the Milky Way runs diagonally across the sky, and the constellation of *Vela* straddles the meridian. Slightly farther south is the constellation of *Carina*, with, to the west, brilliant *Canopus* (α Carinae), which lies below the constellation of *Puppis*, which is itself between Vela and *Canis Major* in the west.

Crux (the Southern Cross) is southeast of Carina and the two principal stars of Centaurus, *Rigil Kentaurus* (α Centauri) and *Hadar* (β Centauri). The *Large Magellanic Cloud* (LMC) lies west of these stars, on the other side of the meridian.

Did you know?
Easter Sunday is calculated to occur after the Full Moon following
March 21. Now that astronomers know the phases of the Moon
decades or hundreds of years in advance, you might think that
knowing the date of Easter would be easy. But it is not as easy as
that. The Christian religious festival is calculated using a lunar
calendar: the ecclesiastical calendar. In this, the 'months' alternate
between 30 and 29 days. To astronomers, the time between New
Moon and New Moon (known as the synodic month) averages
29.530 587 981 days (to take it to nine decimal places). That is just
slightly more than twenty-nine and a half days, so 'months' that
alternate between 29 and 30 days do give a reasonable agreement.
(But not always, see the description of the term 'Black Moon' on
page 62.) In the ecclesiastical calendar, Full Moon is always taken to
be on the 14th day of the lunar month, reckoned in local time. So
the ecclesiastical calendar may get out of step with the astronomical
calendar, in which Full Moon is defined as occurring at a specific
date and time (in Co-ordinated Universal Time, UTC). What is
more, to astronomers, the exact date and time of Full Moon apply
worldwide. In 2021, the religious Easter Sunday is on March 28
(at Full Moon), whereas if it were calculated astronomically it
would be on April 4 at Last Quarter.

April

April – Introduction

April in 2021 is a quiet month astronomically, with few major events. The event that is most likely to come to the notice of the general public is the fact that on April 27, Full Moon occurs extremely close to lunar perigee, when the Moon is closest to the Earth. (Its distance on April 27 is 357,378 km). Full Moon is at 03:32 UT and perigee at 15:22 UT, just under 12 hours later. Although there is one closer perigee in 2021 (on December 4, when the distance is 356,794 km), this Full Moon is the nearest to perigee in 2021 and will almost certainly be announced in the media as a 'supermoon', although as described in the item on the 'Size of the Moon' on pages 94–95, it will not be particularly striking to the naked eye.

There is one occultation of **Mars**, but that may be seen from just a limited area of southeast Asia. (It is described on page 102.)

However, there are no less than three meteor showers active in April, all of which begin in the middle of the month. The first of these, is relatively modest, and is the **Lyrids** (often known as the April Lyrids). In 2021, these begin on April 13 and continue until about April 29, with a peak on the night of April 22–23. Although

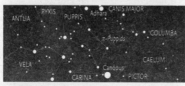

The charts show the location of the Lyrids radiant (top), the π-Puppids radiant (middle) and the η-Aquariids radiant (bottom).

the general hourly rate is not particularly high (about 18–20 meteors per hour), the meteors are fast and some leave persistent trains that may be seen long after the meteor has passed. In 1982, there was a short-lived peak, when the hourly rate reached 90, but it is generally much lower. In 2021, the maximum occurs when the Moon is waxing gibbous, aged 11–12 days, so conditions are not particularly favourable. The radiant is near the star *Vega*, close to the border with *Hercules*. The parent comet for the particles in the meteoroid stream is the non-periodic comet C/1861 G1 (Thatcher).

There is a southern meteor shower, the *π-Puppids*, that begins in April. This shower starts to be active on April 14 and lasts until April 27, with maximum on the night of April 23–24. The shower was unknown until 1972. The rate is rather variable, reaching a maximum of about 40 meteors per hour in 1977 and 1983, and it is difficult to predict whether many meteors will be seen. As with the Lyrids, maximum occurs when the Moon is waxing gibbous. The parent comet is 26P/Grigg-Skjellerup. A more significant shower, the *η-Aquariids*, begins on April 18 and continues well into May (May 27) with maximum on May 6. This shower, like the Orionids of October, is related to debris left in orbit behind the famous comet 1P/Halley. This is a more predictable shower, and the maximum rate is about 40 meteors per hour. The radiant is close to the 'Y'-shaped asterism known as the 'Water Jar' in *Aquarius*.

The Planets

Mercury is very close to the Sun throughout the month, and is thus invisible. It reaches superior conjunction on the far side of the Sun on April 19. Although *Venus* passed superior conjunction on March 26, it remains too close to the Sun, in the evening sky, to be seen, even though it is bright at mag. -3.9. *Mars* fades slightly from mag. 1.3 to 1.5, as it moves from *Taurus* into *Gemini* towards the end of the month. *Jupiter*, initially in *Capricornus*, moves slowly eastwards in the morning sky, and enters *Aquarius* at the very end of the month. It fades very slightly (from mag. 2.1 to mag. 2.2) over the course of the month. *Saturn* is also in the morning sky in Capricornus, and brightens a little (from mag. 0.8 to mag. 0.7). *Uranus* is in *Aries*, and comes to superior conjunction with the Sun on April 30. It is actually at mag. 5.9 throughout April. *Neptune*, in Aquarius, also (like Saturn) brightens very slightly, increasing from mag. 8.0 to mag. 7.9 over the month.

Sunrise and sunset

City	Date	Sunrise	Sunset
Buenos Aires, Argentina			
	Apr .01	10:06	21:48
	Apr. 30	10:29	21:12
Cape Town, South Africa			
	Apr. 01	04:58	16:41
	Apr. 30	05:20	16:06
London, UK			
	Apr. 01	05:36	18:35
	Apr. 30	04:34	19:23
Los Angeles, USA			
	Apr. 01	13:40	02:14
	Apr. 30	13:04	02:36
Nairobi, Kenya			
	Apr. 01	03:34	15:39
	Apr. 30	03:28	15:32
Sydney, Australia			
	Apr. 01	20:08	07:50
	Apr. 30	20:30	07:16
Tokyo, Japan			
	Apr. 01	20:26	09:02
	Apr. 30	19:49	09:26
Washington, DC, USA			
	Apr. 01	10:53	23:32
	Apr. 30	10:11	00:00
Wellington, New Zealand			
	Apr. 01	18:37	06:13
	Apr. 30	19:08	05:29

NB: the times given are in Universal Time (UT)

The Moon's phases and ages

Northern hemisphere

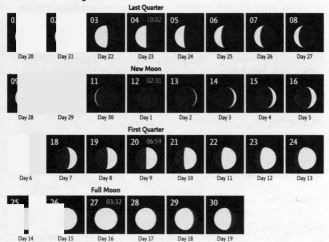

Southern hemisphere

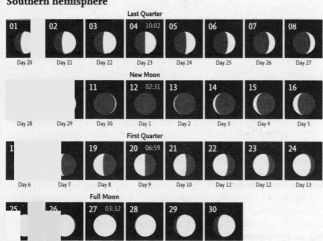

The Moon

On April 1, the waning gibbous Moon is north of **Antares**. On April 6–7, a waning crescent, it passes south of **Saturn** and **Jupiter** in the morning sky. On April 16–17, the waxing crescent Moon passes the **Pleiades** cluster (M45), orange **Aldebaran** and red **Mars**. It passes 0.1° north of Mars on April 17, and there is an occultation, but the end (reappearance) only is visible from southern Asia. (This occultation is described in detail on page 102.) On April 19, the Moon (a day before First Quarter) passes 3.2° south of **Pollux** in **Gemini**. The Moon is 6.5° north of **Spica** in **Virgo** on April 26 and 4.8° north of **Antares** in **Scorpius** on April 29. Full Moon is on April 27, and the Moon is then near perigee, which occurs about 12 hours later. Although one closer perigee occurs later in the year (December 4) this is the closest in time that Full Moon falls near perigee in 2021, so this is likely to be described as a 'supermoon' by the media (see below).

Pink Moon
The Full Moon in April is often known as the Pink Moon. The term derives from the pink flowers, particularly the plants known as phlox, that are prominent in early spring. Other names for this Full Moon are 'Sprouting Green Moon' (from the vibrant green shoots that appear in spring), 'Fish Moon' (because it becomes possible to catch fish again, once the rivers were no longer frozen), and 'Hare Moon' (hares are active during this month). Old-World terms were 'Egg Moon' – eggs were (and are) associated with Easter, and in particular, 'Pascal Moon' (because it was used to calculate the date of Easter – see page 88).

The size of the Moon
A term that has become popular in recent years is 'supermoon'. The apparent size of the Moon varies naturally as a result of its elliptical orbit, relative to the Earth. Every month its distance alters from perigee (closest) to apogee (farthest). Two lunar perigees are separated by one 'anomalistic' month, which is just over 27.5 days (and slightly longer than one lunar orbit at just over 27.3 days.) The dates of both extremes in the distance to the Moon are given in the monthly calendars. The actual distance at perigee and apogee varies throughout the year. The range of distances is 356,400–370,400 km (perigee) and 404,000–406,700 km (apogee). These distances are measured between the centre of the Earth and the centre of the

Moon. Standing on the surface of the Earth, we are closer to the surface of the Moon by about 8000 km.

These changes in distance mean that the apparent size of the Moon alters continuously. However, the change in size is not readily apparent to the eye. Only photographs, taken at carefully arranged times, will show a difference when images at perigee and apogee are compared. When a full moon happens to occur at perigee, it is popularly termed a 'supermoon'. The display is not quite as dramatic as the name suggests, but such a full moon sometimes appears brighter than average. Another effect is seen when the Moon is close to the horizon, either rising or setting, and appears extremely large. This is known as the 'Moon Illusion', which is described on page 146. The Moon is not really larger, but the impression may be very striking.

Surveyor 3, a robotic unit, landed on the Moon on 20 April 1967. It was later visited on 19 November 1969, during the *Apollo 12* mission.

A

These images of First and Last Quarter Moon (top) and Full Moon (bottom) show that the difference in size of the Moon at perigee and apogee is actually very small. (It is about 14 per cent, at most.) In both cases, the apogee image is on the right.

Calendar for April

01	21:19	Antares 4.9°S of Moon
04	10:02	Last Quarter
06	08:29	Saturn 4.0°N of Moon
07	07:17	Jupiter 4.4°N of Moon
09	10:44	Neptune 4.3°N of Moon
11	06:01	Mercury 3.0°N of Moon
12	02:31	New Moon
12	09:48	Venus 2.9°N of Moon
13–29		April Lyrid meteor shower
13	11:41	Uranus 2.5°N of Moon
14–27		π-Puppid meteor shower
14	17:46	Moon at apogee (406,119 km)
16	04:41	Aldebaran 5.5°S of Moon
17	12:08	Mars 0.1°N of Moon
18–May 27		η-Aquariid meteor shower
19	01:49	Mercury at superior conjunction
19	18:51	Pollux 3.2°N of Moon
20	06:59	First Quarter
22		April Lyrid shower maximum
22	10:18	Regulus 4.9°S of Moon
23–24		π-Puppid shower maximum
26	03:13	Spica 6.5°S of Moon
26	09:00 *	Mercury 1.3°N of Venus
27	03:32	Full Moon
27	15:22	Moon at perigee (357,378 km)
29	06:37	Antares 4.8°S of Moon
30	19:54	Uranus at conjunction with the Sun

These objects are close together for an extended period around this time.

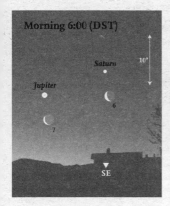

April 6–7 · *The waning crescent Moon passes Saturn and Jupiter low in the southeast (as seen from central USA).*

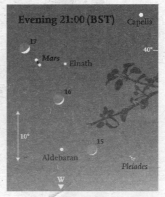

April 15–17 · *The waxing crescent Moon passes Aldebaran, Elnath (β Tau) and Mars, high in the west (as seen from London).*

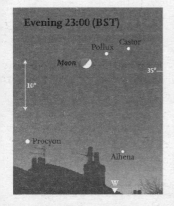

April 19 · *The Moon lines up with Castor and Pollux. Procyon and Alhena (γ Gem) are closer to the horizon (as seen from London).*

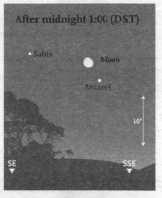

April 29 · *After midnight, the Moon is in the southeast, in the company of Antares and Sabik (η Oph). As seen from central USA.*

A

April – Looking North

Ursa Major is still high overhead for most northern-hemisphere observers, with **Boötes** and bright **Arcturus** (α Boötis), the star indicated by the 'arc' of the 'tail' of Ursa Major, to its northeast. The small circlet of **Corona Borealis**, with its single bright star of **Alphecca** (α CrB) is still farther to the east. **Cassiopeia** has now swung round, and is almost on the meridian to the north, below **Polaris** and northwest of **Ursa Minor**. **Cepheus** is beginning to climb higher and the head of **Draco** is almost due east of Polaris, but slightly farther south than the two 'Guards' in Ursa Minor, **Kochab** (β UMi) and **Pherkad** (γ UMi). On the other side of the meridian to the northwest lies the constellation of **Auriga**, with bright **Capella**. The inconspicuous constellation of **Camelopardalis** lies between Polaris, Auriga and

On 13 April 1970, a ruptured oxygen tank caused the abortion of the **Apollo 13** mission to be the third Moon landing. The crew returned safely to Earth on April 17.

Perseus. The stars of the Milky Way run through Auriga, Perseus, and Cassiopeia towards the much more densely populated regions in **Cygnus** and farther south. Nearly the whole of Cygnus is visible above the horizon in the northeast, where the small constellation of Lyra, with the bright star **Vega**, is clearly seen with **Hercules** above it.

The constellation of Perseus is beginning to descend into the northwest, following the constellation of **Andromeda**, which, for most observers, has now disappeared below the northwestern horizon.

The zodiacal constellation of Virgo is clearly visible, although Spica (α Virginis) is the only really bright star in the constellation.

April – Looking South

The constellation of *Leo* is still conspicuous in the south, although now it is *Denebola* (β Leonis) rather than *Regulus* that is close to the meridian. *Boötes* and *Arcturus* (α Boötis), the brightest star in the northern hemisphere, are high in the southeast. Beyond Leo, along the ecliptic to the east, the zodiacal constellation of *Virgo* is clearly visible, although *Spica* (α Virginis) is the only really bright star in the constellation (see page 99).

Farther south, *Crux* is now nearly 'upright' and close to the meridian, with *Rigil Kentaurus* and *Hadar* (α and β Centauri, respectively), conspicuous to its southeast. Rigil Kentaurus is a multiple system, with two stars fairly easily visible with a small telescope. A third star in the system, known as *Proxima Centauri*, is much fainter, and is actually the closest star to the Solar System at a distance of about 4.3 light-years. In recent years, it has been found to host the closest known exoplanet, which orbits Proxima every 11.2 days.

On 21 April 1972, *Apollo 16* landed in the Descartes highlands on the Moon.

The stars of *Vela* and *Carina* lie to the west of Crux and the meridian, while much farther south, and out to the west is *Canopus* (α Carinae) the second-brightest star in the whole sky after Sirius in Canis Major. Even farther south, and right on the horizon for people at 30°S (roughly the latitude of Sydney in Australia) is *Achernar* (α Eridani), the main star in the long, winding constellation of *Eridanus* (the River), which begins near the star *Rigel* in *Orion*, far to the north.

The constellation of Crux, the Southern Cross, with its two brightest stars, Acrux (α) and Mimosa (β). A little farther east (left) are Rigil Kentaurus and Hadar (α and β Centauri, respectively).

Occultation of Mars

There is an occultation of Mars by the Moon on April 17. This event will be visible only from Indochina (Cambodia, Laos, Vietnam), Thailand, Malaysia and parts of Indonesia. This occultation of Mars lasts about an hour. The exact times of disappearance and reappearance (which are at the Moon's dark and illuminated limbs, respectively) naturally vary depending on the observer's location on Earth. At Singapore, for example, disappearance is at 13:32:6.9 UT and reappearance at 14:33:31.9 UT. The Moon's phase is a waxing crescent (it is day six of the lunation), so the disappearance at the dark limb would be readily detectable, but the reappearance at the bright limb (when visible at all, depending on the observer's location) will be much more difficult to observe.

There are actually three occultations of Mars in 2021, but two of those (on December 3 and 31) are not readily visible. One disappearance (on December 3) is visible for a few minutes in a tiny region of Siberia, and the other (on December 31) occurs primarily in narrow areas over the far Southern Ocean.

The occultation of Mars may be seen from the area enclosed by the solid lines. Although the occultation begins earlier, it cannot be seen from the area of the Earth enclosed by the dotted lines.

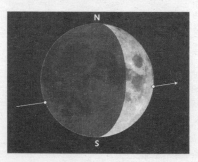

The occultation of Mars as seen from Singapore, with disappearance at 13:33 UT and reappearance one hour later at 14:33 UT.

May

May – Introduction

The most significant astronomical event in May is the lunar eclipse on May 26. As with all lunar eclipses, this is visible over a wide area of the world, although, because maximum eclipse occurs over the Pacific Ocean, the eclipse will not be visible (even partially) from Europe and Africa in particular. The eclipse is described in more detail on pages 108–109.

There is one particularly important meteor shower that is active during May. This is the *η-Aquariid* shower, which began in April, but reaches its maximum on May 6. This is when the Moon is a waning crescent, several days before New Moon, giving reasonably favourable conditions for observations. This shower is produced by debris shed by comet 1P/Halley and shares this feature with the *Orionid* meteor shower in October. These two showers are the only known instance where the Earth regularly encounters, twice in any one year, the stream of particles from a comet.

The η-Aquariids are not particularly favourably placed for northern-hemisphere observers, because the radiant is near the celestial equator, near the 'Water Jar' in Aquarius, well below the horizon until late in the night (around dawn). However, meteors may still be seen in the eastern sky even when the radiant is below the horizon. Although the hourly rate is not particularly high (about 40–50 per hour), as many as a quarter of the meteors leave persistent trains. There is a radiant map for the *η-Aquariids* on page 90.

In late May 2020, it was announced that two new meteor streams, apparently associated with long-period comets, had been discovered. The first of these, the *γ-Piscid Austrinids*, seemed to peak early in the morning of May 15 and may well be an annual shower. The second shower, (the *σ-Phoenicids*) appeared to peak slightly later on the same date. The data suggest that the rate for this shower may vary considerably from year to year. Both showers are known only from low numbers of meteors (15 and 14, respectively).

On 18 May 1969, in a dress rehearsal for Apollo 11, the crewed *Apollo 10* spacecraft, with three astronauts, descended to just 14.3 km of the surface of the Moon.

The Planets

Mercury begins the month at mag. -1.1, but fades to mag. 2.8 at the end. At greatest elongation east (22°) in the evening sky on May 17 it is mag. 0.3. **Venus** (which passed superior conjunction on the far side of the Sun in March), becomes visible for a short period low down on the horizon in the evening sky at about mag. -3.9. **Mars** (at mag. 1.6–1.7) travels across the constellation of **Gemini** in the course of the month. **Jupiter** is fairly bright (mag. -2.2 to -2.4) and is in **Aquarius**. **Saturn** (mag. 0.7–0.6) is in **Capricornus**, and begins retrograde motion, moving eastwards, on May 24. **Uranus** is in **Aries** at mag. 5.9, and **Neptune** in Aquarius at mag. 7.9.

On 5 May 2018, NASA's **Insight** spaceprobe landed successfully at Elysium Planitia on Mars and is currently operational.

M

In May, the constellation of Centaurus is well placed for observing from the tropical regions and farther south. It contains the largest and finest globular star cluster in the sky, Omega Centauri.

Sunrise and sunset

City	Date	Sunrise	Sunset
Buenos Aires, Argentina			
	May 01	10:30	21:11
	May 31	10:51	20:51
Cape Town, South Africa			
	May 01	05:21	16:05
	May 31	05:42	15:45
London, UK			
	May 01	04:32	19:25
	May 31	03:50	20:08
Los Angeles, USA			
	May 01	13:03	02:37
	May 31	12:43	02:59
Nairobi, Kenya			
	May 01	03:28	15:32
	May 31	03:29	15:32
Sydney, Australia			
	May 01	20:30	07:15
	May 31	20:52	06:55
Tokyo, Japan			
	May 01	19:48	09:27
	May 31	19:27	09:51
Washington, DC, USA			
	May 01	10:10	00:00
	May 31	09:45	00:27
Wellington, New Zealand			
	May 01	19:09	05:28
	May 31	19:37	05:01

NB: the times given are in Universal Time (UT)

The Moon's phases and ages

Northern hemisphere

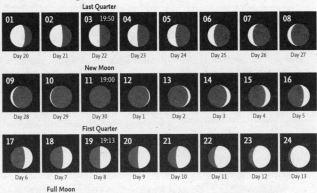

		Last Quarter					
01	02	03 19:50	04	05	06	07	08
Day 20	Day 21	Day 22	Day 23	Day 24	Day 25	Day 26	Day 27

		New Moon					
09	10	11 19:00	12	13	14	15	16
Day 28	Day 29	Day 30	Day 1	Day 2	Day 3	Day 4	Day 5

		First Quarter					
17	18	19 19:13	20	21	22	23	24
Day 6	Day 7	Day 8	Day 9	Day 10	Day 11	Day 12	Day 13

Full Moon						
25	26 11:14	27	28	29	30	31
Day 14	Day 15	Day 16	Day 17	Day 18	Day 19	Day 20

M

Southern hemisphere

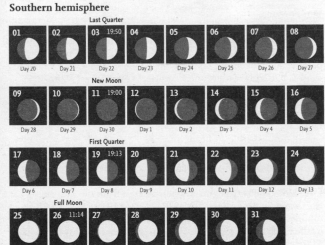

		Last Quarter					
01	02	03 19:50	04	05	06	07	08
Day 20	Day 21	Day 22	Day 23	Day 24	Day 25	Day 26	Day 27

		New Moon					
09	10	11 19:00	12	13	14	15	16
Day 28	Day 29	Day 30	Day 1	Day 2	Day 3	Day 4	Day 5

		First Quarter					
17	18	19 19:13	20	21	22	23	24
Day 6	Day 7	Day 8	Day 9	Day 10	Day 11	Day 12	Day 13

Full Moon						
25	26 11:14	27	28	29	30	31
Day 14	Day 15	Day 16	Day 17	Day 18	Day 19	Day 20

The Moon

On May 3, just before Last Quarter, the Moon is 4.2° south of *Saturn* in *Capricornus*. One day later it passes a similar distance (4.6°) south of *Jupiter* in *Aquarius*. On May 11, the Moon comes to its most distant apogee of the year, when its distance is 406,512 km. This is, however, at New Moon, when the Moon is, effectively, invisible.

On 30 May 1966, the *Luna 10* spaceprobe, the first object to enter orbit around the Moon, stopped transmission after 460 lunar orbits.

One day later, on May 12–13, as a narrow waxing crescent, it passes *Venus*, *Aldebaran* (α Tauri) and then *Mercury*, when the planet is about mag -0.2. A few days later (on May 15–17) the Moon is north of *Mars* in *Gemini* and then 3.1° south of *Pollux* in that constellation. Two days later, on May 19, at First Quarter, the Moon passes between *Regulus* and *Algieba* (γ Leonis) in *Leo*. As waxing gibbous, on 23 May, the Moon is 6.5° north of *Spica* in *Virgo*, and on 26 May (at Full Moon) there is a total lunar eclipse, visible from the Americas, the Pacific, Australia and parts of Asia. This is described more fully on the opposite page. By May 30, the Moon is again 4.2° south of Saturn in Capricornus as it was at the beginning of the month.

Flower Moon

The Full Moon of May (this year on May 26) is sometimes known as the 'Flower Moon' or 'Blossom Moon' among the Chipewa and Objibwe of the Great-Lakes region. This term derives, of course, because so many flowers bloom during the month. Other names for this particular Full Moon from North America are 'Corn-Planting Moon', 'Moon when ice is breaking in rivers', 'Moon of the big leaves' and, among the Cheyenne of the Great Plains 'Moon when the horses get fat'. An Old English/Anglo-Saxon term was 'Milk Moon'.

Total lunar eclipse May 26

In the total lunar eclipse on May 26, maximum eclipse occurs over the middle of the Pacific Ocean. Parts of the eclipse are visible from land. The very beginning may be seen from the west coasts of North and South America for a short period before Moonset. The middle of the eclipse occurs at 11:18.5 UT, when the Moon (shown by the black dot) is high over the central Pacific, roughly south of American Samoa and west of the Cook Islands. The whole eclipse may be seen from New Zealand, and all but the earliest phases from Australia, especially southeastern Australia. The total phases (when the Moon is completely within the Earth's shadow) start at 11:10 UT and end at 12:53 UT. They may be seen from the whole of Australia. Part of totality and the end of the eclipse will be visible from Indonesia and eastern Asia. In this eclipse, the Moon grazes the northern edge of the central umbra of the Earth's shadow, but the duration within the umbra is enough to cause this eclipse to be classed as total.

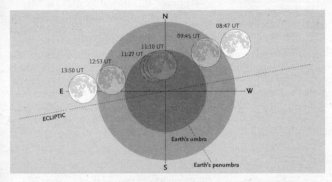

M

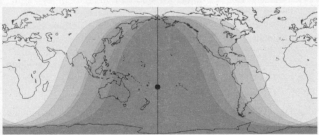

The lines on the map indicate the places where different phases of the eclipse begin. Starting in the east (right) they show when the Moon first contacts the penumbra; when it fully enters the penumbra; first contact with the umbra; and when it fully enters the dark umbra. The lines on the western side are the corresponding phases at the end of the eclipse.

Calendar for May

03	16:58	Saturn 4.2°N of Moon
03	19:50	Last Quarter
04	21:02	Jupiter 4.6°N of Moon
06		η-Aquariid shower maximum
06	17:51	Neptune 4.4°N of Moon
10	21:06	Uranus 2.4°N of Moon
11	19:00	New Moon
11	21:53	Moon at apogee (406,512 km, greatest of year)
12	22:03	Venus 0.7°N of Moon
13	10:48	Aldebaran 5.5°S of Moon
13	17:59	Mercury 2.1°N of Moon
16	04:47	Mars 1.5°S of Moon
17	01:12	Pollux 3.1°N of Moon
17	05:54	Mercury at greatest elongation (22.0°E, mag. 0.3)
17	23:00 *	Venus 5.9°N of Aldebaran
19	17:59	Regulus 5.0°S of Moon
19	19:13	First Quarter
23	13:38	Spica 6.5°S of Moon
26	01:50	Moon at perigee (357,311 km)
26	11:14	Full Moon
26	11:18	Total lunar eclipse (Americas, Pacific, Australia, Asia)
26	17:25	Antares 4.8°S of Moon
29	06:00 *	Mercury 0.4°S of Venus
31	01:18	Saturn 4.2°N of Moon

These objects are close together for an extended period around this time.

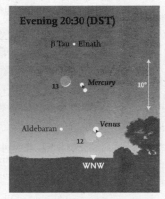

M

May 3–5 • *The Moon is around Last Quarter when it passes Saturn and Jupiter in the southeast (as seen from central USA).*

May 12–13 • *After sunset, the narrow crescent Moon passes Venus, Aldebaran and Mercury (as seen from central USA).*

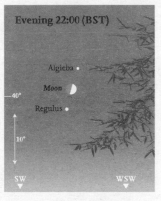

May 15–16 • *The Moon passes Mars, Castor and Pollux in the western sky (as seen from central USA).*

May 19 • *The Moon is between Regulus and Algieba (γ Leo), high in the southwest (as seen from London).*

northwest   north   northeast

May – Looking North

The constellation of *Cassiopeia* is now low over the northern horizon, to the east of the meridian, to observers at mid-northern latitudes. To its west, the southern portions of both *Perseus* and *Auriga* are becoming difficult to observe as they become closer to the horizon, although the *Double Cluster*, between Perseus and Cassiopeia, is still clearly visible.

High above, *Ursa Major* has started to swing round to the west and the stars that form the extended portion towards the south (the 'legs' and 'paws' of the 'bear') are becoming easier to see. *Alkaid* (η Ursae Majoris), at the end of the 'tail', is almost exactly at the zenith for observers at 50° North. The whole of the constellations of *Cepheus*, *Draco*, *Lyra* and *Ursa Minor* are easy to see, together with the inconspicuous constellation of *Camelopardalis*. For those people with the keenest eyesight, they can even make out the long, straggling line of faint stars, forming the constellation of *Lynx* in the northwestern sky, beyond the extended portion of Ursa Major. These faint constellations become difficult to detect during the brighter nights towards the end of the month.

Also high in the sky for northern observers are the constellations of *Boötes*, with bright *Arcturus*, and *Hercules*, with its clearly visible 'Keystone' of four stars, on one side of which is *M13*, the finest globular cluster in the northern sky. The small, but striking arc of stars forming the constellation of *Corona Borealis* lies between Hercules and Boötes. It has just a single bright star, *Alphecca* (α Coronae Borealis).

Much of *Cygnus* is clearly seen in the east, as are two (*Deneb* and *Vega*) of the three stars that form the Summer Triangle. The third star, Altair in Aquila, begins to climb above the horizon late in the night and later in the month. Across the Milky Way, between Cassiopeia and Cygnus, and below Cepheus, is the tiny zig-zag of stars that forms the small, and often ignored, constellation of *Lacerta*.

M13, in the constellation of Hercules is the finest globular cluster in the northern sky.

M

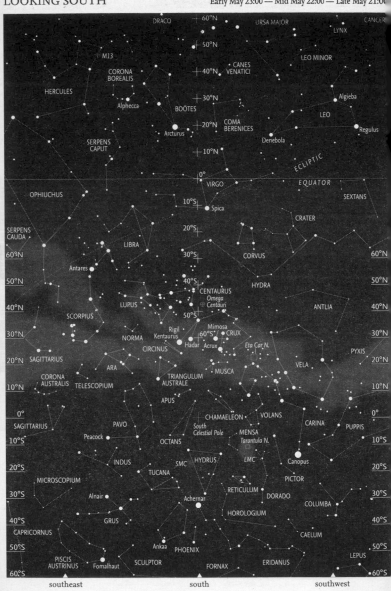

southeast
south
southwest

114

May – Looking South

Spica (α Virginis) in the sprawling constellation of **Virgo** is now close to the meridian. The constellation of **Boötes**, with brilliant **Arcturus** (the brightest star in the northern hemisphere of the sky) is higher above and slightly to the east, clearly visible to observers in the north. Following Virgo across the sky is the rather undistinguished constellation of **Libra** and, following it, and rising in the east, **Scorpius**, with the red supergiant star **Antares** and the distinctive line of stars, running south and ending in the 'sting'. (As Orion sinks in the west, so Scorpius rises in the east. This recalls one of the many legends about Orion: with his being pursued by a scorpion, sent to distract him while fighting.) Above Libra and Scorpius, the large constellation of **Ophiuchus** is beginning to climb higher in the sky. It lies between the two portions of **Serpens** (the only constellation to be divided into two parts). **Serpens Caput** (the Head of the Serpent) lies west of Ophiuchus, between it and Boötes, and **Serpens Cauda** (the Tail of the Serpent) is to the east, between Ophiuchus and the Milky Way.

Farther south, between Scorpius and the two brightest stars in **Centaurus** (**Rigil Kentaurus** and **Hadar**), lies the constellation of **Lupus**, just east of the meridian and lying along the Milky Way. **Crux** is now more-or-less 'upright', with the small constellation of **Musca** below it. Below the bright pair of stars in Centaurus (α and β Centauri) is the constellation of **Triangulum Australe**, a much larger and more striking constellation than its counterpart (Triangulum) in the north. Lying between Rigil Kentaurus and Triangulum Australe is the very tiny and indistinct constellation of **Circinus**.

The stars of **Vela** and **Carina** are becoming lower in the southwest, following the constellation of **Puppis** and brilliant **Canopus** (α Carinae) down towards the horizon. Puppis and the **Large Magellanic Cloud** (LMC) are close to the horizon for anyone at the latitude of 30°S (about the latitude of Sydney in Australia), where **Achernar** (α Eridani) is now too low to be seen.

There are several, small, relatively faint constellations in this part of the sky. The most distinct is probably **Pavo**, with its single bright star (α Pavonis), known as **Peacock**. Other constellations are **Apus**, **Chamaeleon**, **Octans** (which actually includes the South Celestial Pole), **Mensa** and **Volans**.

M

June

June – Introduction

The most important astronomical event in June 2021 is the annular solar eclipse on June 10. Annular eclipses occur only when the Moon is near apogee, at its most distant from the Earth, when its disc does not appear large enough to cover the whole disc of the Sun, leaving an annulus (a ring) uncovered. In this case, the Moon is at apogee on June 8, slightly more than two full days before the eclipse. So the Moon is slightly closer to the Earth, and the annular ring of the Sun is slightly narrower than it would be in an extreme case. This eclipse is notable for the fact that the Moon's shadow nearly misses the Earth and falls in the remote Arctic. Indeed, the path of totality actually crosses the North Pole. Full details of the eclipse and a diagram showing the path of the shadow are given on pages 122–123.

Northern observers may find the time around the summer solstice frustrating, because twilight tends to persist throughout the night. There is one compensation, however: it is at this time of year that displays of noctilucent clouds take place. These highly distinctive clouds, usually particularly visible in the middle of the night, in the direction of the North Pole, are described in detail on page 130–132.

For observers in the southern hemisphere, it is, of course, the time of the winter solstice, and the dark nights make it the ideal time for learning the many faint constellations around the South Celestial Pole. These southern circumpolar constellations are described on pages 28–31.

For both northern and southern observers, there are no significant meteor showers active in June.

On 5 June 2018, the **New Horizons** spaceprobe was successfully 'awakened' from its long hibernation, before the encounter with Arrokoth (then unofficially called Ultima Thule).

The Planets

Mercury is too close to the Sun to be visible. It passes inferior conjunction on June 11. *Venus* is also close to the Sun, but may perhaps be glimpsed in the evening twilight at the end of the month. On June 22, it is south of *Pollux* in *Gemini*, and in the same field as *Mars*. At Last Quarter on June 2 Mars is 5.4°S of Pollux. During the month, the planet moves from Gemini into *Cancer*. *Jupiter* is in *Aquarius* at mag. -2.4 to -2.6 and begins retrograde motion (moving westwards) on June 21. *Saturn* is mag. 0.6–0.4 and is also moving retrograde in *Capricornus*. *Uranus* is mag. 5.9–5.8 and is in *Aries* for the whole month, while *Neptune* (mag. 7.9) is in *Aquarius*, close to the border with *Pisces*. It begins retrograde motion on June 27.

Alphecca

Arcturus

In June, the constellations of Boötes (with Arcturus) and Corona Borealis (with Alphecca) are well placed for observation. Arcturus has an orange tint and is the brightest star in the northern celestial hemisphere.

Sunrise and sunset

City	Date	Sunrise	Sunset
Buenos Aires, Argentina			
	Jun. 01	10:52	20:51
	Jun. 30	11:01	20:53
Cape Town, South Africa			
	Jun. 01	05:43	15:45
	Jun. 30	05:52	15:48
London, UK			
	Jun. 01	03:49	20:09
	Jun. 30	03:47	20:22
Los Angeles, USA			
	Jun. 01	12:43	02:59
	Jun. 30	12:45	03:09
Nairobi, Kenya			
	Jun. 01	03:29	15:32
	Jun. 30	03:35	15:38
Sydney, Australia			
	Jun. 01	20:52	06:54
	Jun. 30	21:01	06:57
Tokyo, Japan			
	Jun. 01	19:26	09:51
	Jun. 30	19:29	10:01
Washington, DC, USA			
	Jun. 01	09:45	00:27
	Jun. 30	09:46	00:38
Wellington, New Zealand			
	Jun. 01	19:38	05:00
	Jun. 30	19:48	05:01

NB: the times given are in Universal Time (UT)

The Moon's phases and ages

Northern hemisphere

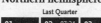

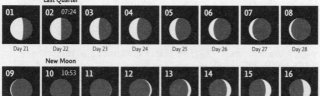

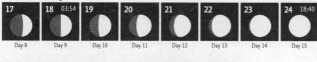

Southern hemisphere

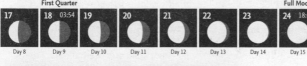

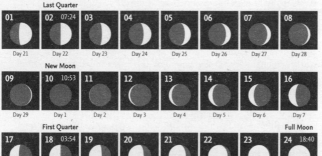

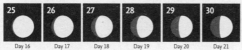

J

121

The Moon

At the beginning of the month, on June 1, one day before Last Quarter, the Moon is 4.6° south of *Jupiter* (mag. -2.4) in *Aquarius*. On June 10, there is the annular solar eclipse, described in detail below and on the facing page, and fully visible just from Arctic Canada and northwestern Greenland. On June 12–13, the Moon passes *Venus* (at mag. -3.8) and *Pollux* in *Gemini* and then moves 2.8° north of the much fainter *Mars* (mag. 1.8). On June 15, the Moon is 5.0° north of *Regulus* in *Leo*, and by June 19 it is close to *Spica* in *Virgo*. On June 27, it passes, first 4.0° south of *Saturn* (mag. 0.4) in *Capricornus*, and then, the next day it is 4.5° south of Jupiter in Aquarius, similar to its position at the beginning of the month.

Strawberry Moon

June has always been noted for strawberries, and this appears in the name of the Full Moon this month (on June 24 in 2021). This name was used on both sides of the Atlantic. The Choctaw tribe of southeastern America had different names for Full Moons that occurred in early or late June. In early June it was 'Moon of the peach' and in late June is was 'Moon of the crane'. Other names are 'Hot Moon' and 'Rose Moon' and in the Old World, 'Mead Moon'.

Annular solar eclipse

On June 10 there is an annular solar eclipse, when the Moon is too far from the Earth to completely cover the disc of the Sun. The Moon will be at apogee (its farthest distance from the Earth) on June 8, just two days before the eclipse. This eclipse occurs at such a high northern latitude that few people will be able to witness the total phases, when just a ring of light is visible around the Moon. The actual ground track where the annular phases will be visible begins in Arctic Canada, sweeps across northwestern Greenland, then across the Arctic Ocean (it actually crosses the North Pole), with the final phases visible in the far corner of northeastern Siberia. The width of the path within which the annular eclipse will be visible is very narrow, being only 527 km across. Partial phases of the eclipse, will, of course, be visible over a much wider area, including most of Europe, except the south of Italy and the Balkans. The start of the eclipse will be visible immediately after sunrise from the northeastern states of the USA and southeastern Canada (Nova Scotia and Newfoundland, in particular).

Maximum eclipse will occur at 10:41:51 UT over northwestern Greenland, not far from the Thule Air Base that is operated by the United States Air Force. The altitude of the Sun (and Moon) will then be just 23.3°. The duration of maximum eclipse will be 3 minutes 51.2 seconds.

As mentioned on page 10, viewing any solar eclipse, even the partial phases, should be undertaken with proper eclipse glasses, by projection, or (if using any optical aid) only with proper solar filters. Although for some observers the Sun will rise eclipsed, even then, staring at the Sun for a long time without proper protection is not advisable.

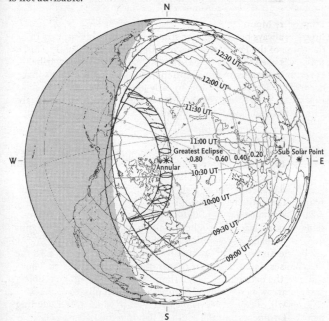

The annular eclipse of June 10. The lines of the 'teardrops' at beginning and end indicate (left to right) the end, mid-eclipse, and beginning of eclipse at sunrise (on left) and sunset (on right).

Calendar for June

01	09:00	Jupiter 4.6°N of Moon
02	07:24	Last Quarter
02	14:00 *	Mars 5.4°S of Pollux
03	01:06	Neptune 4.5°N of Moon
07	06:16	Uranus 2.3°N of Moon
08	02:27	Moon at apogee (406,228 km)
09	16:51	Aldebaran 5.5°S of Moon
10	10:41	Annular solar eclipse (Arctic Canada, Greenland)
10	10:53	New Moon
10	13:09	Mercury 4.0°S of Moon
11	01:13	Mercury at inferior conjunction
12	06:42	Venus 1.5°S of Moon
13	06:52	Pollux 3.1°N of Moon
13	19:52	Mars 2.8°S of Moon
15	23:59	Regulus 5.0°S of Moon
18	03:54	First Quarter
19	22:09	Spica 6.5°S of Moon
21	03:32	June solstice
22	15:00 *	Venus 5.3°S of Pollux
23	03:56	Antares 4.8°S of Moon
23	09:55	Moon at perigee (359,956 km)
24	18:40	Full Moon
27	09:27	Saturn 4.0°N of Moon
28	18:42	Jupiter 4.5°N of Moon
30	09:09	Neptune 4.4°N of Moon

These objects are close together for an extended period around this time.

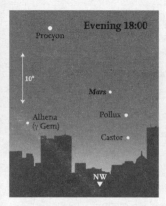

June 2 • *Mars with Castor and Pollux low in the northwest. Mars is now slightly fainter than Castor (as seen from Sydney).*

June 12–13 • *The narrow crescent Moon passes Castor, Pollux and Mars. Venus is close to the horizon (as seen from London).*

J

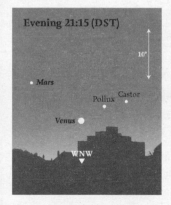

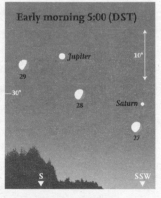

June 22 • *Venus with Castor and Pollux, close to the horizon. Mars is a little farther to the west (as seen from central USA).*

June 27–29 • *The Moon passes Saturn and Jupiter in the early morning, almost due south (as seen from central USA).*

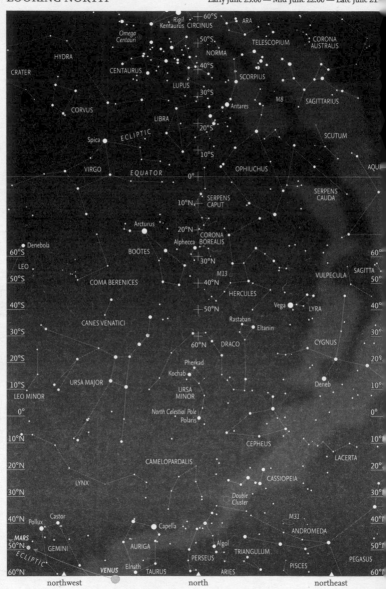

northwest   north   northeast

June – Looking North

For observers in the northern hemisphere, the time around the summer solstice (June 21) is frustrating for observing, because a form of twilight persists throughout the night. The sky remains so light that most faint stars and constellations are invisible unless conditions, such as the lack of light pollution, are particularly favourable. Even the highly distinctive set of seven stars forming the asterism of the 'Plough' (or 'Big Dipper') in **Ursa Major** may be difficult to see. The constellation has now swung round to the west, but the fainter stars of the constellation, lying even farther to the west, are harder to distinguish. The constellation of **Boötes** with bright **Arcturus** (α Boötis) is high overhead for northern observers, and best seen when facing south. The sprawling constellation of **Hercules** lies between **Lyra** and Boötes.

The four stars forming the 'head' of **Draco** lie east of the meridian, although only the brightest, **Eltanin** (γ Draconis) and, possibly **Rastaban** (β Draconis) are clearly visible. Farther north, in **Ursa Minor**, are **Polaris** itself and the two 'Guards', **Kochab** (β UMi) and **Pherkad** (γ UMi).

For observers around 50°N, the southern portions of the constellation of **Auriga** are partially lost on the northern horizon, although bright **Capella** (α Aurigae) should still be visible. Much of the neighbouring, fainter constellation of **Perseus** is difficult to make out. The two main stars of **Gemini**, **Castor** (α Geminorum) and **Pollux** (β Geminorum), the star closest to the ecliptic, and occasionally occulted by the Moon, are low on the northwestern horizon. Somewhat higher in the sky, and northeast of the meridian is **Cassiopeia** with **Cepheus** above it.

Two of the stars forming the angles of the distinctive Summer Triangle, **Deneb** (α Cygni) in **Cygnus** and **Vega** (α Lyrae) in Lyra are clearly visible as, for much of the night, is the third star, Altair (α Aquilae) in **Aquila**. Deneb lies in the Milky Way, at the beginning of the dark Great Rift that runs down the constellation and where obscuring dust prevents us from seeing the dense star clouds of the Milky Way itself.

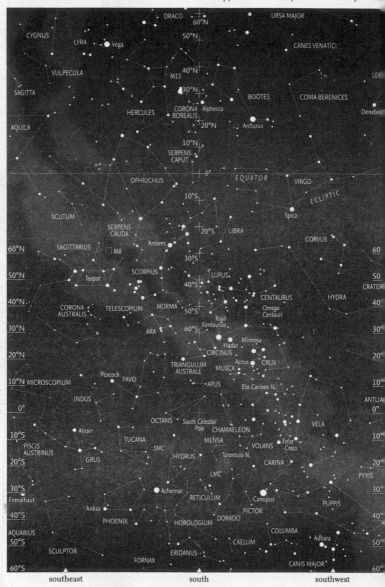

southeast — south — southwest

June – Looking South

The inconspicuous constellation of *Libra* is on the meridian, but the sky is really dominated by the striking constellation of *Scorpius*, slightly to the east. These two were once a single constellation, of course, until the 'claws' of the scorpion were formed into the constellation of The Balance. Although it is rather low for most northern observers, to those farther south, fiery red *Antares* (α Scorpii) is a major beacon and the whole constellation has a very significant presence in the winter skies. The constellation of *Virgo*, and bright *Spica* is now well to the east, along the ecliptic.

At this time of the year, the next constellation in the zodiac, *Sagittarius*, is becoming clearly seen, having risen in the east. The main body of the constellation forms the asterism known as 'The Teapot', but the constellation also has a long, curving chain of faint stars to the south, rather similar to the long line of stars below Scorpius, although curving in the opposite direction. This chain of stars partially encloses the small constellation of *Corona Australis*, although this constellation is not as well-formed as its northern counterpart (Corona Borealis). Rising in the east and following Sagittarius into the sky is the constellation of *Capricornus*. To the west of the 'sting' of Scorpius is the rather untidy tangle of stars that forms the constellation of *Lupus* and part of *Centaurus*.

On 13 June 2006, on its way to the outer Solar System, the *New Horizons* spaceprobe made a fly-by of minor planet 132524 APL, returning images and data.

The small constellation of *Crux* is now well south of the two brightest stars of Centaurus, *Rigil Kentaurus* and *Hadar*. The constellations of *Vela* and *Carina* and the *'False Cross'* are plunging towards the horizon. For observers at the latitude of Sydney in Australia, both *Canopus* (α Carinae) and *Achernar* (α Eridani) are so low that they are often invisible through absorption at the horizon.

The South Celestial Pole in *Octans* is surrounded by faint constellations: Octans itself, *Chamaeleon*, *Volans*, *Mensa*, *Hydrus*, *Tucana*, and *Indus*. Only *Pavo*, to the northeast and *Grus*, to the southeast, are slightly more distinct. However, this area also includes the *Large Magellanic Cloud* (LMC), the largest satellite galaxy to our own. Most of the LMC lies in the constellation of *Dorado*, although some extends into the neighbouring constellation of Mensa.

Noctilucent clouds

June sees the beginning of the season in which noctilucent clouds (NLC) are seen in the northern hemisphere. On rare occasions they may become visible at the very end of May, and the season is effectively centred on the summer solstice in June, so they may also be seen in early July. Because of the lack of land masses in the southern hemisphere, reports of NLC there are much rarer. They are most often seen from the research stations on the Antarctic Peninsula but have been observed from New Zealand, Tasmania and Patagonia.

Noctilucent clouds are bright clouds seen in the sky during the middle of the night. They tend to be observed between latitudes of 50° to 65° in both the northern and southern hemispheres and appear in the general direction of the relevant pole. Closer to the pole, the sky does not get dark enough for the clouds to be visible. They are often 'electric-blue' in colour, although they tend to take on yellowish tints both early in the night and towards dawn.

On 2 June 1966, *Surveyor 1* landed in Oceanus Procellarum, the first NASA spaceprobe to soft land on the Moon.

They are the very highest clouds in the atmosphere, occurring at altitudes of about 82 km. This is towards the upper boundary of the atmospheric layer known as the mesosphere, and they are far above 'normal' clouds, which occur in the lowermost layer of the atmosphere (the troposphere), and are generally no higher than 20 km (and often much lower). They are visible because they are so high that they remain illuminated by sunlight, even when the Sun is below the observer's horizon and the ground and lower clouds are in darkness. For the observer it is astronomical twilight (see page 238).

Noctilucent clouds pose a number of problems. They are known to consist of ice particles. (They have been sampled by sounding rockets.) But for an ice crystal to form, it requires a nucleus of a solid particle of a very particular shape. Where do those particles come from? In addition, ice crystals are, of course, frozen water. Where does the water come from? It is hardly possible for water vapour to rise through the atmosphere from the surface, because it is normally trapped below the temperature inversion (where temperature increases with height) that is always present in the atmosphere at an altitude of about 20 km or less, and which is known as the tropopause. There are occasional breaks in that tropopause, however, which may allow water vapour to reach the upper atmosphere. For many years, the best guess was that the water came 'from outside' in the form of cometary particles. More recently, it has generally been accepted that the water probably derives from the breakdown of methane gas, which is able to migrate upwards from lower levels. Because the concentration of water vapour is so low, the clouds form only at exceptionally low temperatures of around -120°C. Temperatures are lowest, counterintuitively, in summer, when there is upwelling in the atmosphere with resultant cooling.

And what about the solid particles on which the ice crystals form? (These are known as glaciation nuclei.) Where do these come from? One theory was that they consisted of material ejected by major volcanic eruptions. But no known volcanic eruption has ever been violent enough to send particles – even tiny particles – that high in the atmosphere. Another theory was that ice crystals could form around clusters of ions, created by cosmic rays. The general view nowadays is that the glaciation nuclei arrive in the form of micrometeorites from space (in other words, as 'space dust').

Noctilucent clouds seem to be becoming more frequent, but it is not known whether this is a real increase in the frequency of

A bright display of noctilucent clouds, photographed on the night of 30–31 May 2020 from Elgin in Scotland. A strong display for so early in the season.

their formation, or whether there are now more observers who are aware of their existence. Although they occur in both hemispheres, the majority of reports come from the northern hemisphere, where there are not only more observers, but also more land areas from which the clouds may be seen. Although there is a satellite dedicated to the study of NLC: the AIM satellite (Aeronomy of Ice in the Mesosphere), amateur astronomers still play an important part in monitoring the occurrence and appearance of the clouds. The clouds exhibit a whole range of structures and these are known by various names, such as 'billows', 'veils', etc. However, the clouds are actually in the form of a thin sheet of ice crystals, and the structure is actually apparent, created by undulations in the sheet of crystals. The clouds appear 'thicker' where the observer's line of sight passes through a greater length of cloud. When the clouds are overhead, they are so thin that they become invisible.

July

July – Introduction

The month of July saw the highly successful Apollo 15 mission to the Moon in 1971, and this is described in more detail on page 139.

There are no less than four regular meteor showers in July. The first of these is the *α-Capricornids*, which is a moderately long shower that begins early in the month (about July 2) and continues until around August 14. They come to maximum, such as it is, on July 30. Unfortunately, their rate is very low, no more than about 5 meteors per hour. In 2021, maximum occurs when the Moon is waning gibbous, one day before Last Quarter, so the conditions are not very favourable. Despite the low rate, however, the shower often produces bright fireballs. The parent body is a rather obscure comet known as 169P/NEAT. This was discovered by NASA's Near-Earth Asteroid Tracking (NEAT) programme, which ran from 1995 until 2007, and was extremely successful in discovering minor planets, having detected over 40,000 objects, together with a few comets.

The second shower active in June is the *Southern δ-Aquariids*. These begin in mid-month (on July 13) and continue until August 24. Their maximum, like the α-Capricornids, is on July 30, with a waning gibbous Moon. Their hourly rate is somewhat higher, but is still generally below 25 meteors per hour. The parent body is uncertain, but believed to be comet 96P/Macholz.

There is a southern shower, the *Piscis Austrinids*, another rather long shower, which begins on July 14 and continues until August 27. Like the α-Capricornids, the hourly rate is low, around 5 meteors per hour, with maximum on July 28. The parent body is currently unknown.

By far the most important shower that begins in July is that of the *Perseids*. These begin around July 16, but have a strong maximum on August 12–13, so are more appropriately described next month.

Minor plant *(6) Hebe* comes to opposition on July 17 in Aquila (see facing page).

The *Cassini* spaceprobe entered orbit around Saturn on 1 July 2004. It continued observations until 15 September 2017, when it was deliberately de-orbited and destroyed.

The Planets

On July 4, *Mercury* is at greatest western elongation (21.6°), but is low in the morning sky. It rapidly approaches the Sun as it moves towards superior conjunction in August. In the evening sky, *Venus* is bright (at mag. -3.9) but also very low. It moves from *Cancer* into *Leo*, passing *Mars* in the middle of the month. Mars, which is already in Leo, is much fainter at mag. 1.8. It is also low in the evening sky. *Jupiter*, which brightens slightly from mag. -2.6 to mag. -2.8 over the month, is slowly retrograding in *Aquarius*. *Saturn* brightens very slightly (mag. 0.4 to 0.2) and is also retrograding slowly, but in the constellation of *Capricornus*. Of the remaining planets, *Uranus* (mag. 5.8) has a slow direct motion in *Aries*, and *Neptune* (mag. 7.8) is another planet that is retrograding slowly, in *Aquarius*, near the border with *Pisces*. On July 17, the minor planet *(6) Hebe* is at opposition in Aquila (see the charts below).

A finder chart for the position of minor planet (6) Hebe, at its opposition on July 17 (mag. 8.4). The grey area is shown in more detail on the chart below, that shows stars down to magnitude 9.5.

Sunrise and sunset

City	Date	Sunrise	Sunset
Buenos Aires, Argentina			
	Jul. 01	11:01	20:54
	Jul. 31	10:48	21:12
Cape Town, South Africa			
	Jul. 01	05:52	15:48
	Jul. 31	05:39	16:06
London, UK			
	Jul. 01	03:48	20:22
	Jul. 31	04:23	19:51
Los Angeles, USA			
	Jul. 01	12:46	03:09
	Jul. 31	13:04	02:56
Nairobi, Kenya			
	Jul. 01	03:35	15:38
	Jul. 31	03:37	15:41
Sydney, Australia			
	Jul. 01	21:01	06:57
	Jul. 31	20:48	07:15
Tokyo, Japan			
	Jul. 01	19:29	10:01
	Jul. 31	19:49	09:46
Washington, DC, USA			
	Jul. 01	09:47	00:38
	Jul. 31	10:09	00:21
Wellington, New Zealand			
	Jul. 01	19:48	05:02
	Jul. 31	19:29	05:25

NB: the times given are in Universal Time (UT)

The Moon's phases and ages

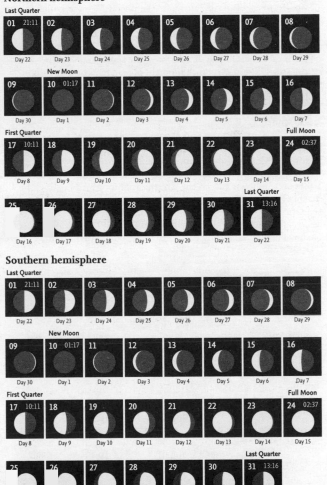

Northern hemisphere

Last Quarter

01 21:11	02	03	04	05	06	07	08
Day 22	Day 23	Day 24	Day 25	Day 26	Day 27	Day 28	Day 29

New Moon

09	10 01:17	11	12	13	14	15	16
Day 30	Day 1	Day 2	Day 3	Day 4	Day 5	Day 6	Day 7

First Quarter / **Full Moon**

17 10:11	18	19	20	21	22	23	24 02:37
Day 8	Day 9	Day 10	Day 11	Day 12	Day 13	Day 14	Day 15

Last Quarter

25	26	27	28	29	30	31 13:16
Day 16	Day 17	Day 18	Day 19	Day 20	Day 21	Day 22

Southern hemisphere

Last Quarter

01 21:11	02	03	04	05	06	07	08
Day 22	Day 23	Day 24	Day 25	Day 26	Day 27	Day 28	Day 29

New Moon

09	10 01:17	11	12	13	14	15	16
Day 30	Day 1	Day 2	Day 3	Day 4	Day 5	Day 6	Day 7

First Quarter / **Full Moon**

17 10:11	18	19	20	21	22	23	24 02:37
Day 8	Day 9	Day 10	Day 11	Day 12	Day 13	Day 14	Day 15

Last Quarter

25	26	27	28	29	30	31 13:16
Day 16	Day 17	Day 18	Day 19	Day 20	Day 21	Day 22

J

The Moon

On July 6, the Moon is 5.6°N of *Aldebaran* in *Taurus* and, on July 8 passes between *Elnath* (β Tauri) and the planet *Mercury*, which is very low in the morning sky. On July 12, the Moon passes north of *Venus* and *Mars* and then, the next day, July 13, crosses between *Regulus* and *Algieba* (γ Leonis). Then at First Quarter, on July 17, it is 6.4°N of *Spica* in *Virgo*. On July 20, it passes due south and is then 4.7°N of *Antares*. On July 24, at Full Moon, it is just 3.8°S of Saturn, and very low in the south. The next day, July 25, the Moon is southwest of Jupiter, passing 4.2°S of the planet a day later on July 26.

Buck Moon
One of the names for the Full Moon for the month of July is 'Buck Moon'. This term derives from the fact that new antlers grow on the heads of male (buck) deer at this time. There are many other names for this Full Moon: among the Chippewa and Ojibwe of the Great Lakes area, following from the 'Strawberry Moon' of June, it is the 'Raspberry Moon'. Then, among the Arapaho of the Great Plains and the Omaha of the Central Plains it is 'the Moon when the buffalo bellow'. There are yet other terms from the Old English/Anglo-Saxon calendar, including the 'Thunder Moon', 'Wort Moon', and 'Hay Moon'.

Apollo 15

On 2 August 1971, the Apollo 15 lander (code-named *Falcon*) descended to the lunar surface near Hadley Rille in Palus Putredinis (see map on the previous page). This mission was the first to use the Lunar Roving Vehicle (LRV), which allowed excursions to a greater distance than with earlier missions. The two members of the crew who landed were David Scott and James Irwin, while Alfred Worden remained in lunar orbit in the Command Module.

The emphasis of this mission was on geology, and the mission returned a large quantity of specimens for study on Earth. This included the large specimen, known as the Genesis Rock, initially thought to be a portion of the Moon's primordial crust. Later analysis showed that it formed later than the Moon itself, but was still about 4100 million years old (4.1 billion years as against 4.5 billion years for the body of the Moon).

The mission was also notable for the demonstration by Scott, who dropped a falcon feather and a geological hammer, which reached the surface simultaneously, confirming the contention of Galileo Galilei in the seventeenth century, that bodies fall at the same rate under gravity, regardless of their weight. On Earth, of course, if the experiment were repeated, air resistance would greatly affect the falcon feather, causing it to flutter about, and take a long time to reach the ground.

J

The Genesis Rock, an anorthosite rock returned by Apollo 15, was originally believed to represent primitive lunar crust, but is now known to have formed later.

Calendar for July

01	21:11	Last Quarter
02–Aug.14		α-Capricornid meteor shower
04	15:25	Uranus 2.1°N of Moon
04	19:45	Mercury at greatest elongation (21.6°W, mag. 0.4)
05	14:47	Moon at apogee (405,341 km)
05	22:27	Earth at aphelion (1.016729224 AU = 152,100,455 km)
06	23:27	Aldebaran 5.6°S of Moon
08	04:38	Mercury 3.8°S of Moon
10	01:17	New Moon
10	12:58	Pollux 3.2°N of Moon
12	09:08	Venus 3.3°S of Moon
12	10:10	Mars 3.8°S of Moon
13–Aug.24		Southern δ-Aquariid meteor shower
13	05:33	Regulus 4.9°S of Moon
13	07:00 *	Mars 0.5°S of Venus
14–Aug.27		Piscis Austrinid meteor shower
16–Aug.23		Perseid meteor shower
17	04:33	Spica 6.4°S of Moon
17	10:11	First Quarter
17	11:29	Minor Planet (6) Hebe at opposition (mag. 8.4)
20	12:38	Antares 4.7°S of Moon
21	10:24	Moon at perigee (364,520 km)
21	19:00 *	Venus 1.2°N of Regulus
24	02:37	Full Moon
24	16:39	Saturn 3.8°N of Moon
26	01:21	Jupiter 4.2°N of Moon
27	17:44	Neptune 4.2°N of Moon
28		Piscis Austrinid shower maximum
29	16:00 *	Mars 0.7°N of Regulus
30		α-Capricornid shower maximum
30		Southern δ-Aquariid shower maximum
31	13:16	Last Quarter

These objects are close together for an extended period around this time.

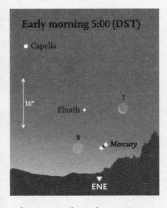

July 7–8 • *In the early morning, the narrow crescent Moon passes between Elnath (β Tau) and Mercury (as seen from central USA).*

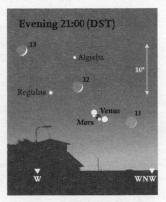

July 11–13 • *The Moon passes Venus and Mars. On July 13, Mars is only 0.5° south of Venus (as seen from central USA).*

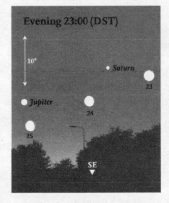

July 23–25 • *The Moon is almost full when it passes Saturn. It is waning gibbous when it reaches Jupiter (as seen from central USA).*

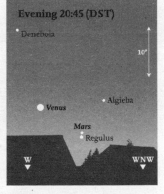

July 29 • *Mars and Regulus are close together with Algieba (γ Leo) and the much brighter Venus nearby (as seen from central USA).*

J

July – Looking North

The brilliant star **Vega** (α Lyrae) is now shining high overhead and the constellations of **Hercules** and **Lyra** are on opposite sides of the meridian, not far from the zenith for observers at 40°N, while it is the head of **Draco** that is near the zenith for observers slightly farther north at 50°N. Beyond Hercules, in the northwestern sky, are the constellations of **Corona Borealis** and **Boötes**, the latter with brilliant, orange-tinted **Arcturus**.

The stars of the Milky Way are now running more-or-less 'vertically', from north to south on the eastern side of the meridian. The constellation of **Cygnus** is 'upside down', high in the sky. For observers at the equator, it is the giant constellation of **Ophiuchus** that is at the zenith, with the two parts of **Serpens** (**Serpens Caput** to the west, and **Serpens Cauda** to the east, among the clouds of the Milky Way). The third star of the Summer Triangle, **Altair** (α Aquilae) in **Aquila** is similarly high in the sky.

Ursa Major is now clearly visible in the northwest, and on the opposite side of the meridian, the constellation of **Cepheus**, with its base in the Milky Way, is at a slightly greater altitude. **Cassiopeia**, the other constellation that, like Ursa Major, is the key to finding one's way around the northern circumpolar constellations, lies in the Milky Way on the opposite side of the North Celestial Pole and **Polaris** in **Ursa Minor**. The faint constellations of **Camelopardalis** and **Lynx** lie to the west and slightly farther south. The chain of faint stars forming Lynx runs 'horizontally' below the outflung stars of Ursa Major. Below Cassiopeia on the other side of the meridian, **Perseus**, with the famous variable star, **Algol**, is beginning to climb higher in the sky and observers at mid-northern latitudes will find that they can now more clearly see **Capella** (α Aurigae) and the northernmost portion of **Auriga**. Observers in the far north (around latitude 60°N) may even occasionally glimpse **Castor** and **Pollux** in **Gemini** peeping above the northern horizon.

On 14 July 2015, the **New Horizons** spaceprobe made a close fly-by of the dwarf planet Pluto, making observations of the planet, its major satellite, Charon, and discovering new, small satellites.

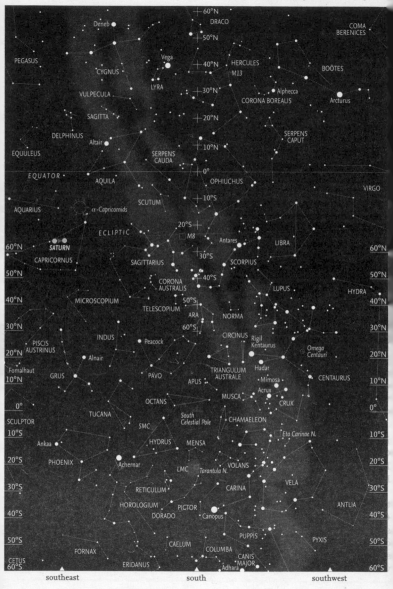

southeast   south   southwest

July – Looking South

The sky looking south is dominated by the two zodiacal constellations of *Scorpius* and *Sagittarius*, which lie on the ecliptic on either side of the meridian, although they are low for observers at most northern latitudes. Bright, red *Antares* (α Scorpii) is very conspicuous, even when it is low in the sky. Not for nothing has it earned the name that means 'Rival of Mars'. The roughly triangular shape of *Capricornus*, the next zodiacal constellation, lies east of Sagittarius and is now clearly visible. Below Scorpius, in the Milky Way, are the small, and often ignored, constellations of *Norma* (which is little more than three stars and easily overlooked) and *Ara*, the latter with a more distinctive shape and brighter stars. The stars of *Lupus* and the outlying stars of *Centaurus* (including the great globular cluster known as *Omega Centauri*) lie farther west.

High above, the constellation of *Ophiuchus* lies at the zenith for observers on the equator, with *Aquila* and bright *Altair* (α Aquilae) to the east. Northwest of Ophiuchus is *Boötes*, the principal star of which is *Arcturus*, which most people see as having an orange tint. Between Arcturus and the meridian is the circlet of stars that forms the constellation of *Corona Borealis*.

South of Scorpius is *Triangulum Australe*, to the east of the two bright stars of Centaurus, *Rigil Kentaurus* and *Hadar*, and lying within the star clouds of the Milky Way. At about the same altitude is the constellation of *Pavo* and, beyond it, *Indus*. Still farther east lies the elongated and somewhat distorted cross-shape that is the constellation of *Grus*. *Piscis Austrinus*, which lies south of Capricornus and Aquarius, has a single bright star, *Fomalhaut* (α Piscis Austrini). This ancient constellation (it was mentioned by Ptolemy, in the second century AD) has now risen in the east.

Crux and the adjoining constellation of *Musca* are now low on the horizon for observers at the equator, and the *Small Magellanic Cloud* (SMC) in *Hydrus* is actually below it. For observers farther south, *Achernar* (α Eridani) is now clearly visible, as is the constellation of *Phoenix* to the east. Only observers in the extreme south, however, will be able to see the whole of *Vela* and *Carina* as well as brilliant *Canopus* (α Carinae).

John Glenn, the first American to orbit the Earth, making three orbits of the planet on 20 February 1962, was born on 18 July 1921. He died on 8 December 2016.

The Moon Illusion

Frequently, when the Moon is seen rising or setting, and it is close to the horizon, it appears absolutely enormous. It seems far larger than when it is high in the sky. In fact, this is an optical illusion – it is known as the 'Moon Illusion'. Our eyes are playing tricks. In reality, the Moon is exactly the same size wherever it is in the sky. Try covering it with your finger, held at arm's length. Any finger is more than large enough to cover the Moon. (In fact, the Moon is about 30 minutes across, and a finger is about twice that, about 1 whole degree.) There have been many attempts over the years to explain the Moon Illusion, but it seems that the brain automatically compares the size of the Moon with the distant horizon, and assumes that it is at the same distance – whereas it is, of course, much farther away.

When the Moon is seen against distant objects, the brain is tricked into thinking it is much larger than its true size.

Thomas Harriott

July 2021 sees the anniversary of the death of the astronomer Thomas Harriot, who was born in Oxford around 1560, graduating from St Mary Hall, Oxford, in 1580. About 1583, he began work with Sir Walter Raleigh and sailed under his auspices to Virginia in 1585. He is noted for his understanding and translations of the Algonquinian language spoken in the Carolinas.

On 26 July 1609 Harriott made the first telescopic observation of the Moon, anticipating the work of the now more famous Galileo Galilei by several months. In about 1613, he drew an extremely accurate map of the Moon, far superior to that produced by Galileo, although unlike the latter, he drew no conclusions about the nature of the surface features. He continued his astronomical observations, and is noted for his work on sunspots in addition to his work in optics (he discovered what later became known as Snell's Law of refraction some 20 years before it was rediscovered by Snellius), mathematics (he introduced a form of algebraic notation and various concepts in spherical trigonometry), as well as navigation.

Harriott died in Threadneedle Street, London, on 2 July 1621, 400 years ago, and was buried in a church destroyed in the Great Fire of 1666, the site of which is now beneath the Bank of England. There is a memorial plaque to him in the Bank's foyer. Although he made some notable contributions to astronomy, his work, including his lunar observations, has been largely overlooked, and there is no lunar crater or even a minor planet named after him.

J

Apollo 11, the first human landing on the Moon, took place on 20 July 1969.

August

August – Introduction

Two planets, *Jupiter* and *Saturn* come to opposition in August. Both are in the constellation of Capricornus, with Saturn at opposition first on August 2 and Jupiter later and farther east, on August 20 (see the chart on the facing page for both planets). Saturn's rings long posed a problem for the identification of the true nature of the planet, and these are described on pages 162–163.

August is the month when one of the best and most reliable of all meteor showers of the year, reaches its peak. This is the *Perseid* shower, which generally begins in the middle of the preceding month (about July 16). The maximum is on the night of August 12–13 in 2021, when the hourly rate may reach 100 meteors per hour. On rare occasions the rate is even higher. In 2021, the peak of the shower is shortly after New Moon, so conditions are favourable. The parent body for the Perseids is Comet 109P/Swift-Tuttle (otherwise known as the Great Comet of 1862). Perseid meteors are fast and a number of the brighter ones leave persistent trains. The shower is also notable for occasional bright fireballs.

Three other meteor showers, although reaching maxima in July, may still exhibit activity in August. These are the *α-Capricornids*, which may persist until August 14, the *Southern δ-Aquariids* (until August 24) and the southern shower, the *Piscis Austrinids*, which may persist until August 27. All these showers are described more fully on pages 14–15.

A brilliant Perseid fireball, streaking alongside the Great Rift in the Milky Way, photographed in 2012 by Jens Hackmann from near Weikersheim in Germany. A few additional, fainter Perseids are also visible in the image.

The Planets

On August 1, **Mercury** comes to superior conjunction on the far side of the Sun, so there is a long period of invisibility. **Venus** is in **Virgo**, but is far too low to be easily visible, although it may be glimpsed very briefly in the evening twilight. **Mars** is in **Leo**, and initially it is close to **Regulus** (α Leonis). It is moving eastwards towards **Virgo**, but is very low in the sky and difficult to observe. **Jupiter** (relatively constant in brightness at mag. -2.8 to -2.9 over the month) is in **Capricornus** and comes to opposition on August 20. **Saturn** (at mag. 0.2) is also in Capricornus and reaches opposition earlier in the month, on August 2. **Uranus** is mag. 5.8–5.7, moving slowly in **Aries**, where it begins retrograde (westwards) motion on August 20. **Neptune** remains in **Aquarius** at mag. 7.9.

On 4 August 2007, NASA's **Phoenix** spaceprobe launched. It landed successfully at Vastitas Borealis on Mars on 25 May 2008

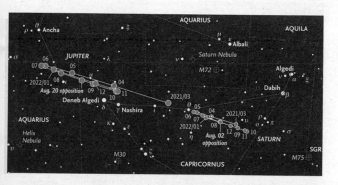

The paths of Jupiter (left) and Saturn (right) from 1 March 2021, to 1 January 2022, in the constellations of Capricornus and Aquarius. The dates of both oppositions are marked. Background stars are shown down to magnitude 6.5.

Sunrise and sunset

City	Date	Sunrise	Sunset
Buenos Aires, Argentina			
	Aug. 01	10:47	21:13
	Aug. 31	10:13	21:34
Cape Town, South Africa			
	Aug. 01	05:39	16:07
	Aug. 31	05:06	16:28
London, UK			
	Aug. 01	04:24	19:49
	Aug. 31	05:12	18:49
Los Angeles, USA			
	Aug. 01	13:05	02:55
	Aug. 31	13:26	02:21
Nairobi, Kenya			
	Aug. 01	03:37	15:41
	Aug. 31	03:31	15:36
Sydney, Australia			
	Aug. 01	20:47	07:16
	Aug. 31	20:14	07:36
Tokyo, Japan			
	Aug. 01	19:50	09:45
	Aug. 31	20:13	09:10
Washington, DC, USA			
	Aug. 01	10:09	00:20
	Aug. 31	10:37	23:40
Wellington, New Zealand			
	Aug. 01	19:28	05:26
	Aug. 31	18:46	05:55

NB: *the times given are in Universal Time (UT)*

The Moon's phases and ages

Northern hemisphere

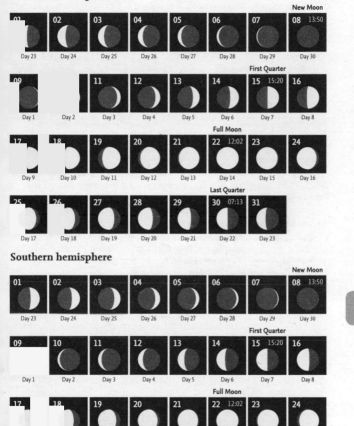

Northern hemisphere

							New Moon
01	02	03	04	05	06	07	08 13:50
Day 23	Day 24	Day 25	Day 26	Day 27	Day 28	Day 29	Day 30

					First Quarter		
09		11	12	13	14	15 15:20	16
Day 1	Day 2	Day 3	Day 4	Day 5	Day 6	Day 7	Day 8

					Full Moon		
17	18	19	20	21	22 12:02	23	24
Day 9	Day 10	Day 11	Day 12	Day 13	Day 14	Day 15	Day 16

				Last Quarter		
25	26	27	28	29	30 07:13	31
Day 17	Day 18	Day 19	Day 20	Day 21	Day 22	Day 23

Southern hemisphere

							New Moon
01	02	03	04	05	06	07	08 13:50
Day 23	Day 24	Day 25	Day 26	Day 27	Day 28	Day 29	Day 30

					First Quarter		
09	10	11	12	13	14	15 15:20	16
Day 1	Day 2	Day 3	Day 4	Day 5	Day 6	Day 7	Day 8

					Full Moon		
17	18	19	20	21	22 12:02	23	24
Day 9	Day 10	Day 11	Day 12	Day 13	Day 14	Day 15	Day 16

				Last Quarter		
25	26	27	28	29	30 07:13	31
Day 17	Day 18	Day 19	Day 20	Day 21	Day 22	Day 23

A

The Moon

The Moon passes north of **Aldebaran** in **Taurus** on August 3. On August 6, as a waning crescent, it is between **Castor** (α Geminorum) and **Alhena** (γ Geminorum), later that day it is 3.1° south of **Pollux**. On August 10, just after New Moon (which is on August 8), the hair-thin waxing crescent passes **Mars**, very low in the evening sky. The next day (August 11), it is 4.3° north of **Venus**. The Moon (as a waxing crescent, two days before First Quarter) passes **Spica** in **Virgo** on August 13 and then is 4.5° north of **Antares** in **Scorpius**, three days later on August 16 (one day after First Quarter). On August 20, the Moon is 3.7° south of **Saturn**, and the next day 4.0° south of **Jupiter**, both in Capricornus. As at the beginning of the month, it passes Aldebaran in Taurus again, on August 30.

Sturgeon Moon

The Algonquin tribes of North America called the Full Moon of August the 'Sturgeon Moon', because of the numerous fish in the lakes where they fished. To the Cree of the Canadian Northern Plains it was the 'Moon when young ducks begin to fly'. Many names refer to the berries ripening at this time: 'berries'; 'black cherries'; and 'chokeberries'. Others refer to the fact that corn is ripening, such as among the Ponca of the Southern Plains: 'Corn is in the silk Moon'. To the Haida in Alaska it was the 'Moon for cedar bark for hats and baskets'. In the ancient Old English/Anglo-Saxon calendar it was sometimes known as 'Barley Moon', 'Fruit Moon' and 'Grain Moon'.

On 25 August 1989, the spaceprobe **Voyager 2** had its encounter with Neptune, the outermost planet visited by that time.

Ernest Mouchez

Ernest Amédée Mouchez was born on 21 August 1821. He entered the École Navale in Paris and after graduation, served in the French Navy, rising through the ranks and eventually becoming contre-amiral in 1878. In 1874, he was in charge of an expedition to successfully observe the transit of Venus on December 9 from Saint Paul Island in the Indian Ocean. In 1878 he succeeded Le Verrier as Director of the Paris Observatory, where he initiated, and carried out numerous improvements to the equipment and functioning of the establishment, including implementing many international links with other astronomical institutions. A crater in the northern region of the Moon is named after him.

Mouchez |

The crater named after Mouchez is near the North Pole of the Moon. Its location is shown on the map (left), and the image below is one obtained by the Lunar Reconnaissance Orbiter, which has carried out detailed mapping of the whole Moon.

A

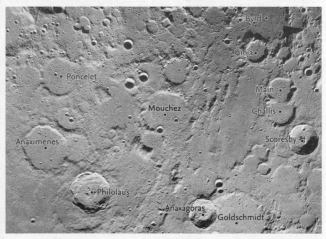

Calendar for August

01	00:29	Uranus 1.8°N of Moon
01	14:07	Mercury at superior conjunction
02	06:14	Saturn at opposition (mag. 0.2)
02	07:35	Moon at apogee (404,410 km)
03	06:53	Aldebaran 5.7°S of Moon
06	20:14	Pollux 3.1°N of Moon
08	13:50	New Moon
09	03:19	Mercury 3.4°S of Moon
09	12:04	Regulus 4.8°S of Moon
10	00:41	Mars 4.3°S of Moon
11	06:59	Venus 4.3°S of Moon
12–13		Perseid shower maximum
13	10:03	Spica 6.1°S of Moon
15	15:20	First Quarter
16	19:06	Antares 4.5°S of Moon
17	09:16	Moon at perigee (369,124 km)
19	04:00 *	Mercury 0.1°S of Mars
20	00:28	Jupiter at opposition (mag. -2.9)
20	22:15	Saturn 3.7°N of Moon
22	04:56	Jupiter 4.0°N of Moon
22	12:02	Full Moon
24	01:55	Neptune 4.0°N of Moon
28–Sep.05		α-Aurigid meteor shower
28	08:57	Uranus 1.5°N of Moon
30	02:22	Moon at apogee (404,100 km)
30	07:13	Last Quarter
30	14:55	Aldebaran 6.0°S of Moon

These objects are close together for an extended period around this time.

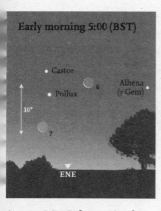

August 6–7 · *Before sunrise, the narrow crescent Moon passes below Castor and Pollux, with Alhena nearby (as seen from London).*

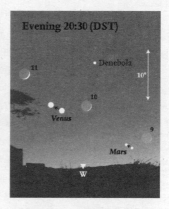

August 9–11 · *The narrow crescent Moon passes Mars and Venus. Mars (mag. 1.8) may not be easy to detect (as seen from central USA).*

August 19 · *Mercury and Mars are close together, separated by only 0.1°. Venus is twenty degrees higher (as seen from Sydney).*

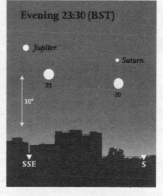

August 20–21 · *The Moon is south of Saturn on August 20. One day later, it is approaching Jupiter (as seen from London).*

A

157

northwest   north   northeast

August – Looking North

The Milky Way is now running up the eastern side of the sky, where *Cassiopeia* may be found within it, high in the northeast. The constellation of *Cepheus*, with the 'base' of the constellation, which is shaped like the gable-end of a house, lies within the edge of the Milky Way, and the whole constellation is 'upside-down' slightly farther north than Cassiopeia. For observers south of the equator, these constellations are close to the horizon and *205* is on the horizon.

For most mid-latitude northern observers, *Perseus* is now clearly visible, as is the northern portion of *Auriga* with brilliant *Capella* (α Aurigae). Perseus straddles a narrow portion of the Milky Way and although not particularly distinct, a significant section of the Milky Way actually runs through Auriga in the direction of *Gemini*. Auriga contains numerous open clusters and a few emission nebulae, but is without globular clusters. Beyond Perseus in the east lies *Andromeda* and the small constellation of *Triangulum*. Slightly farther north than Perseus and on the other side of the meridian is *Ursa Major*, and for most observers even the stars in the southern, extended portion are visible.

For observers at about 40°N, both *Lyra* and *Cygnus*, with their brilliant principal stars, *Vega* and *Deneb*, respectively, are high overhead, near the zenith. The third star marking the other apex of the Summer Triangle, *Altair* in *Aquila*, is much farther to the south, and all three stars are best seen when looking south. The constellation of *Hercules* lies farther west of Lyra and Cygnus, with most of the constellation of *Pegasus* to the east. The tiny, but highly distinctive constellation of *Delphinus* lies between the Milky Way and the outlying stars of Pegasus. Again, these areas are best seen when looking south.

A

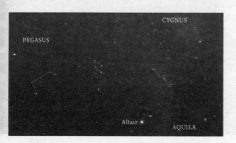

There are four small constellations between Cygnus, Pegasus, and Aquila. They are, from left to right: Equuleus, Delphinus, Sagitta, and Vulpecula.

August – Looking South

The star clouds of the Milky Way dominate the sky at this time of the year. The three stars forming the Summer Triangle, **Deneb** (α Cygni), **Vega** (α Lyrae) and **Altair** (α Aquilae) are clearly visible, as is the Great Rift, running south from near Deneb. The dark Great Rift marks the location of dense clouds of dust that obscure the light from the many stars in the main plane of the Galaxy. Both **Cygnus** and **Aquila** represent birds, and in modern charts these are shown flying 'down' the Milky Way, towards the south. Older charts tended to show Aquila as flying 'across' the Milky Way, in the direction of **Aquarius**, in the east.

For observers at northern latitudes, the constellations of **Hercules** and **Pegasus** are visible, one on each side of the meridian, together with **Delphinus** on the east. Between Cygnus and Aquila, in the Milky Way, lie the two small constellations of **Vulpecula** and **Sagitta**.

The two zodiacal constellations of **Scorpius** and **Sagittarius** are still clearly visible, although becoming rather low for observers at mid-northern latitudes, with even the bright red supergiant star of **Antares** in Scorpius becoming difficult to see. (For observers at 50°N, it is skimming the horizon.) North of Scorpius is the large constellation of **Ophiuchus**, with part of **Serpens**, **Serpens Cauda**, lying in front of the Great Rift.

Capricornus is clearly seen, as is **Aquarius**, the next constellation along the ecliptic. For observers farther south, the curl of stars forming **Corona Australis** lies south of Sagittarius and, farther east, **Piscis Austrinus** with its single bright star, **Fomalhaut** (α Piscis Austrini) is at about the same altitude, with the chain of faint stars that curves south from Sagittarius and the faint constellation of **Microscopium** in between the two. South of Piscis Austrinus is the constellation of **Grus**, with the undistinguished constellation of **Indus** between it and **Pavo**, which is on the meridian.

On 28 August 1993, the **Galileo** spaceprobe flew by the minor planet 243 Ida, obtaining images and data, and discovering Dactyl, the first known satellite of a minor planet.

Farther south, **Lupus**, **Centaurus** and **Crux** are descending towards the horizon, while for observers at 30°S, **Vela** has disappeared, as has most of **Carina**, including brilliant **Canopus**.

Saturn's rings

Although the planet Saturn has been known since antiquity, it was a long time after the discovery of the telescope before the planet's true nature was known. The changing aspect of the planet, caused by the changing inclination of the ring system to the view from Earth, proved to be difficult to understand. When the Earth passes through the ring plane, the rings effectively disappear. (The last time this happened was in 2009.) When Galileo Galilei initially observed the planet through his primitive telescope in 1610, he was confused by the appearance and believed that he was seeing a collection of three bodies orbiting together in a fixed configuration. Although he later saw the planet as a single object, he never realised the reasons for the change in appearance. Other observers (including Pierre Gassendi, whose work is described on page 171) also saw a complex image, with what appeared to be 'ansae' (handles) on each side of the planet. It was a number of years before Christiaan Huygens, using a better telescope in 1655, discovered the rings' true nature as a flat disc surrounding the planet. He published his findings in the form of an anagram: but the solution was not widely known until 1659. It was Huygens who discovered Saturn's largest satellite, Titan, the only satellite in the Solar System to have a dense atmosphere. (ESA's Huygens probe landed successfully on Titan on 14 January 2005.)

On 25 August 1981, the *Voyager 2* spaceprobe got a gravity assist from Saturn on its journey to the outer Solar System.

In 1675, Domenico Cassini noted that Saturn's rings consisted of several individual rings with gaps between them. The widest of these gaps was later named the Cassini Division in his honour. It was 1787 before Pierre Simon Laplace realised that a single disc would not be stable, and that the ring system might consist of a series of solid ringlets. For a number of years it was even suggested that the rings might consist of a fluid. In 1857, James Clerk Maxwell proved conclusively that the rings were neither solid rings nor fluid, but must comprise countless individual particles. It is now accepted that most of these particles consist of water ice, with mere traces of solid, rocky material.

Cassini made numerous studies of Saturn, discovering four satellites. (At the time of writing there are 82 known satellites, 53 of which are named. Apart from the seven larger satellites, 14 of the smaller, named ones are between 10 and 50 km in diameter, and 34 are less than 10 km across.) Because of his contribution to studies of Saturn, the NASA spaceprobe that conducted studies of the planet and its satellites for 13 years from 1 July 2004 was named after him.

It is now realised that the whole system actually consists of innumerable ringlets. However, there are major, effectively permanent, gaps, and there are seven major rings (lettered A to G). The locations of the main rings and the gaps between them are largely governed by gravitational resonances with the main large satellites (and also by a few smaller bodies embedded within the rings).

The origin of Saturn's striking system of rings remains a mystery. For a long time it has been believed, following theoretical calculations, that the ring system formed early in the history of the Solar System. However, recent results from the Cassini spacecraft suggest that their formation was a relatively recent event. For the present, the matter is undecided.

A

Saturn, as imaged by the Hubble Space Telescope on 20 June 2019. The major Cassini Division is clearly seen, as are other narrower gaps in the rings.

September

September – Introduction

For northern observers, the autumnal equinox (the southern spring equinox) occurs on September 22, when the Sun in is Virgo. This is the equinox that is sometimes called the 'Cusp of Virgo'. Technically, the Sun crosses the celestial equator, from north to south at 19:21 UT, when the Earth is at a distance of 1.003773312 astronomical units (AU) or 150,162,279 km.

There is one minor meteor shower that is active in September and one that begins in the month. The first is the **α-Aurigid** shower, which begins on August 28, and continues until September 5. The shower tends to show two peaks, one at the very end of August, and the other (in 2021) on September 1. But the hourly rate is very low, being below 10, and more likely to be just 6, or even less. The meteors, although few, do tend to be bright and are usually easy to photograph. For either date, the Moon is just past Last Quarter (days 23 to 25 of the lunation), so observing conditions are reasonable. The parent body for this meteor stream is comet C/1911 N1 Kiess.

The **Southern Taurid** meteor shower begins on September 10 and is a very long shower, ending about November 20. It comes to a peak in October, so will be more fully described next month. As a slight compensation for the lack of meteor shower activity in September, during the month the number of sporadic meteors reaches its highest rate at any time in the year.

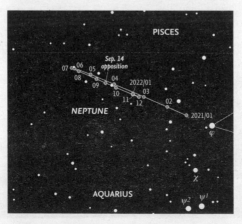

The path of Neptune in 2021. Neptune comes to opposition on September 14. All stars brighter than magnitude 8.5 are shown.

The Planets

Mercury is very close to the Sun and invisible in the evening twilight. It is at greatest elongation east on September 14, but too low to be seen. At the beginning of the month, *Venus* is close to *Spica* in *Virgo*, also close to the Sun. It becomes slightly higher towards the end of the month, but is still deep in the evening twilight. *Mars* is another planet that is too close to the Sun to be seen. *Jupiter* (mag. -2.9 to -2.7 over the month) is still retrograding in *Capricornus* after its opposition on August 20. *Saturn* (mag. 0.3– 0.5) is also retrograding extremely slowly in the same constellation, Capricornus, following its opposition on August 2. *Uranus* (mag. 5.7) is in *Aries*, and *Neptune* (mag. 7.8) reaches opposition in *Aquarius* on September 14 (see the chart on the facing page). Minor planet *(2) Pallas* is at opposition at mag. 8.5 in western *Pisces* on September 11 (see the chart below).

The path of the minor planet (2) Pallas around its opposition on September 11 (mag. 8.5). Background stars are shown down to magnitude 9.5.

Sunrise and sunset

City	Date	Sunrise	Sunset
Buenos Aires, Argentina			
	Sep. 01	10:12	21:35
	Sep. 30	09:31	21:56
Cape Town, South Africa			
	Sep. 01	05:04	16:28
	Sep. 30	04:24	16:48
London, UK			
	Sep. 01	05:14	18:47
	Sep. 30	06:00	17:41
Los Angeles, USA			
	Sep. 01	13:27	02:20
	Sep. 30	13:47	01:40
Nairobi, Kenya			
	Sep. 01	03:30	15:35
	Sep. 30	03:19	15:26
Sydney, Australia			
	Sep. 01	20:12	07:37
	Sep. 30	19:33	07:57
Tokyo, Japan			
	Sep. 01	20:14	09:09
	Sep. 30	20:36	08:27
Washington DC, USA			
	Sep. 01	10:37	23:38
	Sep. 30	11:04	22:52
Wellington, New Zealand			
	Sep. 01	18:45	05:56
	Sep. 30	17:55	06:26

NB: the times given are in Universal Time (UT)

The Moon's phases and ages

Northern hemisphere

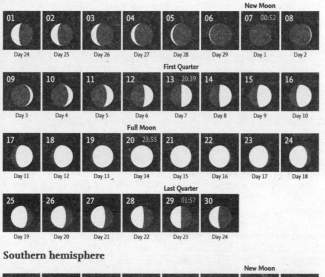

Southern hemisphere

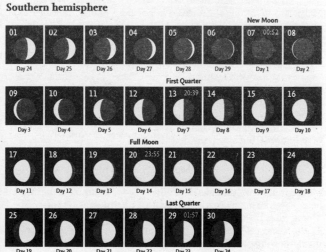

The Moon

On September 3 the Moon passes 3.1° south of **Pollux** in **Gemini**. Two days later, on September 5, the waning crescent Moon is 4.8° north of **Regulus** in the morning sky. By September 9, two days after New Moon, the narrow crescent is 5.9° north of **Spica**, although the star itself will be difficult to detect in the evening twilight. The next day, September 10, the Moon passes 4.1° north of **Venus**.

On 13 September 1959, the spaceprobe **Luna 2** impacted the Moon, the first object to reach the Moon's surface.

Over the days of September 17–18, the Moon passes south of first **Saturn** and then **Jupiter**, both in the constellation of **Capricornus**. On September 25, the Moon passes south of the **Pleiades** cluster and then, on September 26, is 6.2° north of **Aldebaran** in **Taurus**. On September 30, the broad waning crescent Moon, one day after Last Quarter, is 2.8° south of **Pollux** in **Gemini**.

Corn Moon

To the peoples of North America, September was particularly important because corn (maize) was ready for harvest. Many tribes had names for the Full Moon that referred to corn, such as 'Corn Maker Moon' among the Abenaki of northern Maine; 'Middle Moon between harvest and eating corn' to the Algonquin in the Northeast and Great-Lake area, although it was the 'Moon when freeze begins on stream's edge' for the Cheyenne of the Great Plains.

In Europe, the September Full Moon was generally called the 'Harvest Moon', and technically this was the first Full Moon after the autumnal equinox (22 September in 2021). In general, this Full Moon comes in September, but approximately once every three years, Full Moon comes in October, when it is that Full Moon that is known as the 'Harvest Moon'. The term 'Harvest Moon' was the only name for the Full Moon that was determined by the equinox, rather than being specific to any particular month. Other names for this Full Moon, from the Old World, and specifically from the Old English/Anglo-Saxon calendar, were 'Full Corn Moon' – in this case 'corn' meaning wheat or barley – as well as 'Barley Moon'.

On 24 September 1970, **Luna 16** was the first robotic mission to return a sample from the Moon (from Mare Fecunditatis).

Pierre Gassendi

On 12 September 1621, there was a major auroral event. At the time – and indeed, for many years – aurorae were not understood and were regarded with superstition. (Aurorae are described in detail on pages 192–193.) That particular major display was observed by Pierre Gassendi (1592–1655), a priest, philosopher, astronomer and mathematician. He was the first to introduce the term 'aurora borealis' for the phenomenon. Ten years later, he was the first person to observe a transit of Mercury (see page 70) across the Sun, on 7 November 1631, an event that had been predicted by Kepler (see page 222–223). He did attempt to observe the transit of Venus in December that year, but was frustrated by the fact that the event occurred during the night-time in Paris.

Gassendi made many other discoveries and observations, including confirming Pascal's finding that pressure decreased with altitude in the atmosphere. He is commemorated by a prominent (and significant) crater on the Moon, notable for the fractures on its floor, the central peaks and accompanying hills. Minor planet 7179, which was discovered in 1991, is named after him.

The lunar crater Gassendi, on the northern edge of Mare Humorum, has a number of notable features. There are major fractures on the floor and large central peaks caused by the rebound of underlying rock layers following the major impact that created the crater.

S

Calendar for September

01		α-Aurigid shower maximum
03	04:37	Pollux 3.0°N of Moon
05	06:00 *	Venus 1.7°N of Spica
05	20:16	Regulus 4.8°S of Moon
07	00:52	New Moon
07	16:21	Mars 4.2°S of Moon
08	20:19	Mercury 6.5°S of Moon
09	16:32	Spica 5.9°S of Moon
10–Nov.20		Southern Taurid meteor shower
10	02:09	Venus 4.1°S of Moon
11	01:48	Minor planet (2) Pallas at opposition (mag. 8.5)
11	10:03	Moon at perigee (368,461 km)
13	00:31	Antares 4.2°S of Moon
13	20:39	First Quarter
14	04:24	Mercury at greatest elongation (26.8°E, mag. 0.1)
14	09:21	Neptune at opposition (mag. 7.8)
17	02:33	Saturn 3.8°N of Moon
18	06:54	Jupiter 4.0°N of Moon
20	08:45	Neptune 4.0°N of Moon
20	23:55	Full Moon
22	19:21	September equinox
24	16:08	Uranus 1.4°N of Moon
26	21:44	Moon at apogee (404,640 km)
26	22:54	Aldebaran 6.2°S of Moon
29	01:57	Last Quarter
30	13:18	Pollux 2.8°N of Moon

These objects are close together for an extended period around this time.

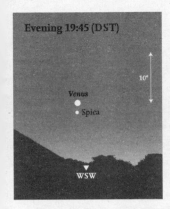

September 5 · *After sunset, Spica is less than two degrees below the much brighter Venus (as seen from central USA).*

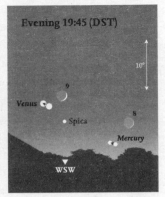

September 8–9 · *After sunset, the narrow crescent Moon passes Mercury, Spica and Venus (as seen from central USA).*

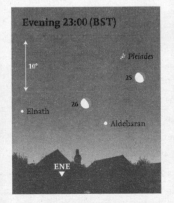

September 25–26 · *The waning gibbous Moon passes the Pleiades and Aldebaran. Elnath (β Tau) is farther north (as seen from London).*

September 30 · *The Moon is in the company of Castor and Pollux, with Alhena (γ Gem) more to the south (as seen from London).*

S

173

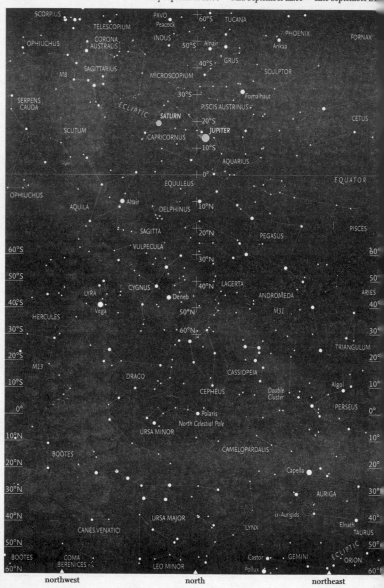

September – Looking North

For mid-northern observers, **Cassiopeia** is high overhead, and **Cepheus** is 'upside-down', apparently hanging from the Milky Way, high above the Pole. **Perseus** is high in the northeast, at about the same altitude as **Ursa Minor** and **Polaris**.

For those same observers, **Ursa Major** is now 'right way up', low in the south, although some of the southernmost stars are lost along or below the horizon. **Auriga**, with bright **Capella**, and the small triangle of the 'Kids' is clearly visible in the northeast. Higher in the sky, above Perseus, the whole of **Andromeda** is visible, together with the small constellation of **Triangulum** and, still higher the Great Square of **Pegasus**, and beyond it, most of the zodiacal constellation of **Pisces**.

The clouds of stars forming the Milky Way are not particularly striking in Auriga and Perseus, but beyond Cassiopeia, and on towards **Cygnus**, they become much denser and easier to see.

Cygnus is high in the northwest and **Deneb**, its principal star, is close to the zenith for observers at 40–50° north, with **Lyra** and bright **Vega**, slightly farther west. Farther towards the south, most of **Hercules** is clearly visible, with the 'Keystone' and the globular cluster M13. The head of **Draco** lies between Hercules and Cepheus and the whole of that constellation is easily seen as it curls around Ursa Minor and the North Celestial Pole.

Observers in the far north may be able to detect **Castor** (α Geminorum) skimming the northern horizon, although brighter **Pollux** (β Geminorum) will be too low to be detected until later in the night.

The constellation of Cassiopeia, with its distinctive 'W' (or 'M') shape, is circumpolar for most northern observers (see pages 26–27).

S

September – Looking South

The three stars forming the apices of the (northern) Summer Triangle, **Deneb** (α Cygni), **Vega** (α Lyrae), and **Altair** (α Aquilae) are prominent in the southwest. The Great Rift is clearly visible, starting near Deneb and running down the centre of the Milky Way towards the centre of the Galaxy in **Sagittarius**, and beyond into **Scorpius**, only petering out in **Centaurus** and **Crux**.

Most of the constellation of **Capricornus** lies just west of the meridian, with **Aquarius** just slightly farther north on the eastern side. The next zodiacal constellation, **Pisces**, is clearly seen, including the prominent asterism, known as the 'Circlet'. Much of the constellation of **Cetus** is visible south of it. South of Aquarius is **Piscis Austrinus**, with its single bright star, **Fomalhaut**, in an otherwise fairly barren area of sky. Still slightly farther south is the undistinguished constellation of **Sculptor**, and, closer to the meridian, the line of stars that forms part of **Grus**.

For observers south of the equator, Sagittarius is high in the west. Below it is the curl of stars that is **Corona Australis**. Below (south of) Sagittarius is the tail and 'sting' of Scorpius. Still farther south is the constellation of **Lupus** and the scattered stars of **Centaurus**, with **Rigil Kentaurus** and **Hadar** (α and β Centauri, respectively). Between the 'sting' of Scorpius and those two bright stars are the small constellations of **Ara** and **Triangulum Australe**.

On 21 September 2003, the **Galileo** spaceprobe was deliberately de-orbited to plunge into the upper atmosphere of Jupiter and be destroyed.

Next comes Crux itself. If the long axis of the cross is extended right across the sky over the largely empty area of sky around the South Celestial Pole, it points in the general direction of **Achernar** (α Eridani). Between Achernar and Ara, west of the meridian, lies the constellation of **Pavo** with it sole bright star **Peacock** (α Pavonis). To the south, the brightest star of the southern hemisphere, **Canopus** (α Carinae) is hugging the horizon for observers at the latitude of Sydney in Australia.

S

Did you know?

Have you heard of **Peter Pan discs**? The investigation of what are known as protoplanetary discs around young stars is of basic importance to studies of how stars and planets form. Such discs are thought to be fundamental to the formation of stars themselves as material (primarily gas) from the inner edge of a disc accretes into a single body at the centre, eventually forming a star. But protoplanetary discs may be gas- and dust-rich, providing material for the formation of actual planets.

On 3 September 1976, the **Viking 2** spaceprobe landed successfully at Utopia Planitia on Mars.

A large number of such discs of material are known, and, of these, some shows distinct signs of the existence of planetary bodies. In a few, planets (or at least large objects) have been detected, and in others there are major gaps, suggesting that the material has either collected into a significant body, or else that some (invisible) body has created a gap.

The consensus of opinion was that protoplanetary discs had a lifetime of a few million years (perhaps 10 million years at most), but in recent years much older discs have been discovered, with lifetimes some 5 to 10 times longer. These discs have been termed 'Peter Pan discs', because, like the fictional boy, they 'never grew up'. The very latest research suggests that such discs occur only when stars form in isolated 'lonely' environments, unlike most stars which are believed to form in clusters, usually amounting to some 100,000 stars.

Peter Pan discs appear to form around low-mass stars, which have been found to host considerable numbers of planets. Study of such discs is likely to prove of considerable importance in understanding planetary formation.

October

October – Introduction

October may be said to mark the beginning of the 'auroral season', because nights are dark enough for aurorae to be readily visible. It is also the month when shipping lines begin to offer 'auroral cruises'. Other than aurorae (described on pages 192–193), October is notable for the fact that there are no less than four meteor showers active during the month.

The first to become active is the *Southern Taurid* shower which starts on September 10 and comes to its peak on October 10, when the Moon is a waxing crescent (Day 5 of the lunation). The shower is not, however, particularly spectacular, with a rate of just 5 meteors per hour (or fewer). The parent body is comet 2P/Encke. This comet is also the parent for the *Northern Taurid* shower, which begins on October 20 and lasts until December 10. Maximum is on November 12, but it also has a low hourly rate of 5 meteors or fewer.

Far more significant are the other two showers: the *Orionids* and the *Draconids*. The Orionids begin on October 1 and continue until November 6, with their peak on October 21. (In 2021, this is one day after Full Moon, so conditions are extremely poor.) The rate is not particularly high, being about 25 meteors per hour. The shower, however, is significant because it is one of the two showers associated with comet 1P/Halley, the other being the η-Aquariid shower, active in April–May.

Finally, the Draconids (also called the October Draconids to distinguish them from other showers with a radiant in Draco) are known to be a periodic shower, which produced very striking meteor 'storms' in 1933 and 1946 with incredible rates that have been estimated at 500 meteors per hour, or even more. Rates in other years have been far less, with normal maximum rates of about 20 meteors per hour. There have been other outbursts, such as one in 2011 when the rate reached approximately 300 per hour, and also in 2012 when a large number of very faint meteors (about 1000 per hour) were detected by radar. A high rate of visual meteors occurred again in 2018, with a peak of about 150 meteors per hour, lasting about 4 hours. Maximum for 2021 is predicted to be on the night of October 8–9, but the likely rate is quite unknown. The parent body is the comet 21P/Giacobini-Zinner (which is why the stream was formerly known

The first artificial satellite, *Sputnik 1*, was launched by the Soviet Union on 4 October 1957.

as the Giacobinids). To add to the peculiarities of the shower, Draconid meteors are exceptionally slow-moving. The radiant is very close to (and almost within) the quadrilateral of stars forming the 'Head' of Draco. In 2021, the Moon is a thin waxing crescent 2–3 days after New Moon, so will not cause any significant interference. Observations are best carried out early in the night, when the radiant is highest in the sky.

The Planets

Mercury passes inferior conjunction on October 9 and is mag. -0.6 at greatest western elongation (18.4°) on October 25. *Venus* is close to *Antares* in *Scorpius* for several days, when both are low in the eastern sky, and passes 1.5° due north of the star on October 16. It is very bright (mag. -4.5) when at its greatest possible eastern elongation (47°) on October 29. *Mars* is at superior conjunction on October 8. *Jupiter* (mag. -2.5 to -2.7) is initially retrograding slowly in *Capricornus* but begins direct motion on October 30. *Saturn*, (mag. 0.5–0.6), is also in Capricornus, and reverts to direct motion on October 19. *Uranus* (mag. 5.7) is retrograding in *Aries*. *Neptune* (mag. 7.8) continues its retrograde motion in *Aquarius*.

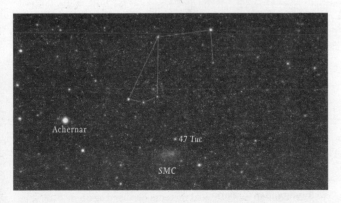

For observers at 20° north or farther south, the small constellation of Tucana is now visible in the south, almost on the meridian (see chart on page 190). The Small Magellanic Cloud (SMC) and the bright globular cluster 47 Tucanae lie within its borders. Achernar (α Eridani) is farther east.

Sunrise and sunset

City	Date	Sunrise	Sunset
Buenos Aires, Argentina			
	Oct. 01	09:30	21:57
	Oct. 31	08:52	22:22
Cape Town, South Africa			
	Oct. 01	04:23	16:49
	Oct. 31	03:46	17:13
London, UK			
	Oct. 01	06:02	17:38
	Oct. 31	06:53	16:35
Los Angeles, USA			
	Oct. 01	13:48	01:38
	Oct. 31	14:12	01:02
Nairobi, Kenya			
	Oct. 01	03:19	15:26
	Oct. 31	03:12	15:21
Sydney, Australia			
	Oct. 01	19:31	07:58
	Oct. 31	18:55	08:22
Tokyo, Japan			
	Oct. 01	20:36	08:25
	Oct. 31	21:03	07:47
Washington, DC, USA			
	Oct. 01	11:04	22:50
	Oct. 31	11:35	22:08
Wellington, New Zealand			
	Oct. 01	17:54	06:27
	Oct. 31	17:08	07:01

NB: *the times given are in Universal Time (UT)*

The Moon's phases and ages

Northern hemisphere

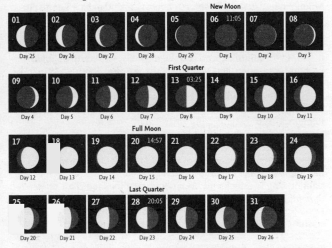

New Moon

01	02	03	04	05	06 11:05	07	08
Day 25	Day 26	Day 27	Day 28	Day 29	Day 1	Day 2	Day 3

First Quarter

09	10	11	12	13 03:25	14	15	16
Day 4	Day 5	Day 6	Day 7	Day 8	Day 9	Day 10	Day 11

Full Moon

17	18	19	20 14:57	21	22	23	24
Day 12	Day 13	Day 14	Day 15	Day 16	Day 17	Day 18	Day 19

Last Quarter

25	26	27	28 20:05	29	30	31
Day 20	Day 21	Day 22	Day 23	Day 24	Day 25	Day 26

Southern hemisphere

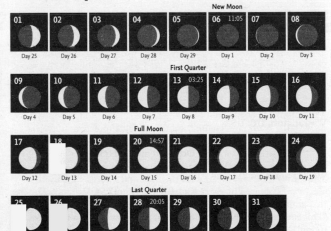

New Moon

01	02	03	04	05	06 11:05	07	08
Day 25	Day 26	Day 27	Day 28	Day 29	Day 1	Day 2	Day 3

First Quarter

09	10	11	12	13 03:25	14	15	16
Day 4	Day 5	Day 6	Day 7	Day 8	Day 9	Day 10	Day 11

Full Moon

17	18	19	20 14:57	21	22	23	24
Day 12	Day 13	Day 14	Day 15	Day 16	Day 17	Day 18	Day 19

Last Quarter

25	26	27	28 20:05	29	30	31
Day 20	Day 21	Day 22	Day 23	Day 24	Day 25	Day 26

O

The Moon

On October 3, three days before New Moon, the waning crescent Moon passes between **Regulus** (α Leonis) and **Algieba** (γ Leonis) in the early morning sky. On October 9, the Moon is 2.9° north of **Venus**, and it passes 4.0° north of **Antares** in **Scorpius**, the next day, October 10, when low in the evening sky. On October 14, the Moon passes 3.9° south of **Saturn**, and the next day, 4.1° south of **Jupiter**, both in **Capricornus**. By October 24, the waning gibbous Moon (Day 19 of the lunation) is 6.4° north of **Aldebaran** in **Taurus**. On October 27, still waning gibbous, the Moon passes 3.0° south of **Pollux** in **Gemini**, and three days later, on October 30, it is 5.1° north of Regulus in **Leo**.

Hunter's Moon

In the northern hemisphere of the Old World, October was the month in which people prepared for the coming winter by hunting, slaughtering livestock and preserving meat for food. This caused the Full Moon in October to become known as the 'Hunter's Moon'. Every three years, however, the first Full Moon after the autumnal equinox fell, not in September, but early in October, when it was also the 'Harvest Moon'. The October Full Moon was also known as the 'Dying Grass Moon' and the 'Blood' or 'Sanguine Moon'. ('Blood Moon' is also the term sometimes applied to the Moon during a lunar eclipse.)

In the New World there was a great variety of names. Many were related to the changes in autumn, such as 'Leaf-falling Moon', 'Falling Leaves Moon' or simply 'Fall Moon'. To the Algonquin of the Northeast and Great Lakes area it was the 'White Frost on Grass Moon'. The Assiniboine of the Northern Plains had a rather different sort of name. To them it was the 'Joins Both Sides Moon'.

The colour of the Moon

As with the term 'supermoon', another North-American term has been widely adopted. This involves the use of the term 'Blue Moon'. Originally, this term was used by meteorologists for the optical phenomenon where the Moon literally appears blue in colour. The same effect may occur with the Sun, and in both cases, was (and still is) caused by particles suspended in the upper atmosphere that scatter the appropriate wavelengths (colours) of sunlight. (This rare event is the actual origin of the phrase 'Once in a blue moon'.) The most common source of such particles in the upper atmosphere, which need to be of a specific size, is a wildfire. On occasions, forest fires in Canada have produced so many particles and spread them so widely downwind that tinted suns and moons have been seen even in Europe. A somewhat similar effect may sometimes occur when volcanoes eject large quantities of ash into the upper atmosphere, although the colours tend to be orange and red, because of a different range of particle sizes. Orange and red colours are often associated, in Europe, with incursions of particles from the North-African deserts. A different range of particle sizes from forest fires (in particular) may give rise to a green coloration of the Moon or Sun, but this occurrence is exceptionally rare.

Increasingly, however, the term 'Blue Moon' came to be applied to the third full moon of the four that occur in a meteorological season or to the second full moon that occurred in a particular calendar month. Nowadays, the latter usage is frequently applied by the media to the second full moon in any month. It is this usage that has crossed the Atlantic to be taken up and publicized by European media.

In reality, of course, the colour of the Moon is very subdued. To the eye, overall, it is simply different shades of grey. There are only lighter and darker areas (the highlands and the maria, respectively). With optical aid, a very skilled observer can detect differences in colour in a few areas, such as a slight reddish tint to part of the floor of the crater Fracastorius, caused by the mineral variations on the Moon's surface.

The colour of the Moon, as seen by the naked eye, is grey – various shades of grey. Only exceptional circumstances, such as forest fires, dust storms or a lunar eclipse cause it to appear any other colour.

O

Calendar for October

01–Nov.06		Orionid meteor shower
03	05:42	Regulus 5.0°S of Moon
06	09:40	Mars 3.6°S of Moon
06	11:05	New Moon
06	17:39	Mercury 6.9°S of Moon
07–11		Draconid meteor shower
07	01:13	Spica 5.8°S of Moon
08	04:01	Mars in conjunction with the Sun
08	17:28	Moon at perigee (363,386 km)
08–09		Draconid shower maximum
09	16:18	Mercury at inferior conjunction
09	18:36	Venus 2.9°S of Moon
10		Southern Taurid shower maximum
10	06:58	Antares 4.0°S of Moon
13	03:25	First Quarter
14	07:08	Saturn 3.9°N of Moon
15	10:03	Jupiter 4.1°N of Moon
16	14:00 *	Venus 1.5°N of Antares
17	13:58	Neptune 4.1°N of Moon
20	06:00 *	Mars 2.8°N of Spica
20	14:57	Full Moon
20–Dec.10		Northern Taurid meteor shower
21	21:40	Uranus 1.3°N of Moon
24	06:17	Aldebaran 6.4°S of Moon
24	15:28	Moon at apogee (405,615 km)
25	05:30	Mercury at greatest elongation (18.4°W, mag. -0.6)
27	21:15	Pollux 3.0°N of Moon
28	20:05	Last Quarter
29	20:52	Venus at greatest elongation (47.0°E, mag. -4.5)
30	15:04	Regulus 5.1°S of Moon

These objects are close together for an extended period around this time.

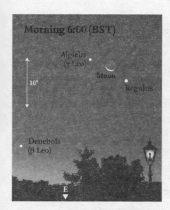

October 3 · *The waning crescent Moon is between Regulus and Algieba. Denebola is close to the horizon (as seen from London).*

October 8–9 · *The crescent Moon passes Venus which is close to Dschubba (δ Sco). As seen from central USA.*

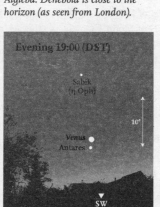

October 16 · *Antares is 1.5° below the much brighter Venus, shortly after sunset (as seen from central USA).*

October 28 · *Castor and Pollux, in the northern sky with the almost Last Quarter Moon (as seen from Sydney).*

October – Looking North

The Milky Way arches across the northern sky, running from **Auriga** in the east to **Cygnus** and **Aquila** in the west. The constellations of **Perseus**, **Cassiopeia**, and **Cepheus** lie along it. For observers in middle northern latitudes, these constellations are high overhead. (To observers in the far north, Cassiopeia is near the zenith.) **Andromeda** is high in the north, beyond Perseus and the other side of the Milky Way. The two small constellations of **Triangulum** and **Aries** are to the south of it. In the northwest, the constellation of **Lyra** lies farther south, clear of the Milky Way.

Between Cassiopeia and Cygnus lies the zig-zag of faint stars that form the small constellation of **Lacerta**, which is often difficult to recognize because it lies across the Milky Way. Two of the stars forming the Summer Triangle, **Deneb** (α Cygni) and **Vega** (α Lyrae) are readily visible, but the third, **Altair** (α Aquilae), is approaching the northwestern horizon. So too, much farther north, is the constellation of **Hercules**. The head of **Draco** is at about the same altitude as **Polaris** in

On 7 October 1959, **Luna 3** returned the first image of the far side of the Moon.

Ursa Minor, as is **Capella** (α Aurigae). The large constellation of **Ursa Major** is now directly beneath the Pole, visible above the horizon to the north. **Castor** and **Pollux** (α and β Geminorum, respectively) are in the northeast, while the northernmost stars of **Boötes** and the circlet of **Corona Borealis** are in the northwest. The brightest star in Boötes, orange-tinted Arcturus, is below the horizon early in the night.

The long chain of faint stars that is the constellation of **Lynx**, runs almost vertically between Ursa Major and **Gemini**, with the other faint constellation of **Camelopardalis** between it and Perseus.

Alnair

For observers at 35°N or farther south, the cross-shaped constellation of Grus is now clearly visible above the southern horizon.

O

southeast   south   southwest

October – Looking South

The Great Square of **Pegasus** is now on the meridian due north, with the constellation of **Andromeda** stretching away from it in the northeast. (The star at the northeastern corner of the Great Square is actually **Alpheratz** (α Andromedae.) The two parts of the zodiacal constellation of **Pisces** are to the south and east of Pegasus. **Aquarius** and the non-zodiacal constellation of **Cetus** are slightly farther south while **Capricornus** is sinking in the west.

South of Aquarius is **Pisces Austrinus** with its single bright star, **Fomalhaut**, while straddling the meridian slightly east of it is the faint constellation of **Sculptor**. Below (south) these two constellations are the roughly cross-shaped constellation of **Grus** with, on the other side of the meridian, **Phoenix**. Just below the constellation of Pheonix is **Achernar** (α Eridani), the bright star that ends the long winding constellation of **Eridanus** that

On 29 October 1991, the **Galileo** spaceprobe encountered the minor planet (951) Gaspra.

actually begins near **Rigel** in **Orion**. Below Achernar is the triangular constellation of **Hydrus**, with the **Small Magellanic Cloud** (SMC) on its northwestern side. At about the same altitude towards the west is the constellation of **Pavo** with **Peacock** (α Pavonis).

The **Large Magellanic Cloud** (LMC) lies southeast of Hydrus, partly in each of the two faint constellations of **Dorado** and **Mensa**. The Milky Way lies in the east, where portions of **Sagittarius** and **Scorpius** are clearly visible. Roughly level with the 'Sting' of Scorpius are the constellations of **Ara** and **Triangulum Australe**. Farther east are the undistinguished constellations of **Apus**, **Chamaeleon** and **Volans**. Even farther to the east is brilliant **Canopus** (α Carinae), the second brightest star in the sky (after Sirius) at mag. -0.6.

For observers at about 30° south (roughly the latitude of Sydney in Australia), the two brightest stars of **Centaurus** (**Rigil Kentaurus** and **Hadar**) are brushing the horizon, while the small constellation of **Crux** is lost below it. Only observers even farther south will be able to see that constellation in full, together with the constellation of **Lupus**, the full extent of Centaurus and the large constellation of **Vela**, with **Puppis** in the southeast.

O

Aurorae

Although aurorae are created by eruptions of material from the Sun and thus occur all year round, they are most readily visible during the darker months of the year. The auroral 'season' may be said to begin (in the northern hemisphere) in October after the brighter nights have passed. The heating of the upper atmosphere that aurorae cause has been suggested as one reason for the lack of sightings of the simultaneous occurrence of aurorae and noctilucent clouds (pages 130–131), although they have been recorded together on rare occasions.

Aurorae have been known from antiquity and for hundreds of years were regarded with superstitious awe. The full name 'aurora borealis' ('northern dawn') was first used by Pierre Gassendi (page 171) who observed a major display on 12 September 1621. The term 'aurora' ('dawn') on its own was first used by Galileo Galilei. (The corresponding phenomenon in the southern hemisphere is, of course, known as the 'aurora australis', but because of the lesser area of land in the south, is generally seen only from the southern tip of South America, Tasmania in southern Australia and New Zealand, as well as from Antarctica.) Serious investigations began at about the same period of the seventeenth century and it was in 1741 that the famous Swedish scientist Anders Celsius, using observations made by his brother-in-law, Olof Hörter, was able to link the occurrence of auroral displays to disturbances of the Earth's magnetic field. Even so, it was not until the early twentieth century that the overall nature of auroral displays were scientifically understood, largely through the work of the Norwegian scientist, Kristian Birkeland.

The Earth is constantly bathed in the solar wind, a stream of plasma (electrons and charged ions). These particles are trapped by the Earth's magnetic field behind the Earth in what is known as the magnetotail. They are accelerated 'up' the tail towards the Earth and, because of the shape of the Earth's magnetic field, cascade into the upper atmosphere around the poles, in the auroral zone, which is typically some 10 to 20° away from the geomagnetic poles and some 3 to 6° wide in latitude. Active displays occur in what are known as the 'auroral ovals', which are displaced towards the night side of the Earth. There, the particles collide with atoms in the upper atmosphere, raising them to higher energies. When the atoms drop back to lower energy levels, they emit the characteristic light seen as an aurora. A display in one hemisphere is often matched by a similar display in the other hemisphere.

Major displays are initiated by activity on the Sun, including flares on the surface. Particularly strong events may cause major changes to the Earth's magnetic field, with disruption to power-distribution systems on the surface, as well as endangering satellites and other sensitive electronic equipment. The very strongest effects occur when there is an event known as a coronal mass ejection (CME) in which highly energetic plasma is released into the solar wind from the Sun's corona. Depending on the location of the release, the surge in plasma follows the lines of the interplanetary magnetic field and may impact the Earth.

The majority of auroral displays occur between altitudes of 90 and 150 km above the Earth's surface, although occasionally they extend to altitudes of 1000 km or more. Even the lowest aurorae are thus above even the highest noctilucent clouds (at altitudes of around 82 km), although it has been suggested that auroral heating may extend to lower altitudes and thus affect the formation of NLC.

The aurora photographed from a ship, north of the Arctic Circle, en route for Narvik in Norway, on 16 March 2019. Photographed by Alan Tough.

O

November

November – Introduction

There is a lunar eclipse on November 19, the second eclipse in 2021. As with the first on May 26, maximum eclipse takes place over the Pacific Ocean, and this time it occurs to the east of the Hawaiian Islands. Details are given on pages 200–201.

There is considerable meteor activity in November. At the beginning of the month, there is the tail of the **Orionid** shower, which began on October 1 and peaked on October 21. The **Northern Taurids**, the second shower associated with Comet 2P/Encke, began on October 20, and peaks on November 12. (This shower continues until December 10.) A minor southern shower, the **Phoenicids**, begins on November 22, and peaks in early December.

But, above all, November is the month of the **Leonids**. They begin on November 5 and continue until November 29, with a peak in 2021 on the night of November 17–18. It is this shower that has been responsible for some of the greatest meteor 'storms', which were particularly notable in 1833, when thousands of meteors per hour, 'rained' down. (There were far too many for accurate figures to be obtained.) The trails of particles left behind the parent comet (Comet 55P/Tempel-Tuttle) are affected by planetary perturbations, particularly by the gravitational effect of Jupiter. Some trails consist of few particles, giving rates of just a few per hour. The denser trails produce the high rates, which tend to recur at intervals of approximately 33 years, often with major activity in the years immediately before and after the peak. There was a major storm in 1966, and storms in 1999, 2001 and 2002, when as many as 3000 meteors per hour were recorded. The rate in 2021 is predicted to be much lower, perhaps just 15 per hour. Leonid meteors are very fast, and often produce very bright fireballs (mag. -1.5 or higher).

The path of minor planet (1) Ceres, around its opposition on November 27. Background stars are shown down to magnitude 8.5.

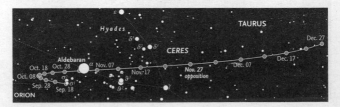

The Planets

Mercury is so close to the Sun that it is invisible this month. It comes to superior conjunction on the far side of the Sun on November 29. **Venus** is in **Sagittarius** and passes south of **Nunki** (σ Sagittarii) on November 19. **Mars** is mag. 1.6 in **Virgo**, and may be glimpsed in the morning twilight. **Jupiter** (mag. -2.5 to -2.3 over the month) is in **Capricornus**, near Deneb Algedi (δ Capricorni). **Saturn** is also in Capricornus, at the other end of the constellation, but much fainter (at mag. 06–0.7). **Uranus** (mag. 5.7) is in **Aries** and reaches opposition on November 4. (See the charts below.) **Neptune** (mag. 7.8) continues its slow retrograde motion in **Aquarius**. Dwarf planet **(1) Ceres** is at opposition at mag. 7.0 in **Taurus**, not far from Aldebaran, on November 27 (see the chart on the facing page).

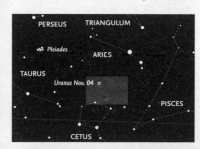

The position of Uranus at its opposition on November 4. The grey area is shown in more detail on the chart below. On that chart, stars down to magnitude 7.5 are shown.

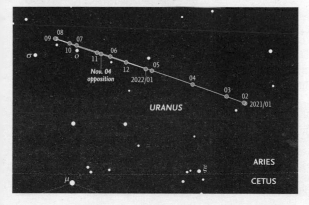

N

Sunrise and sunset

City	Date	Sunrise	Sunset
Buenos Aires, Argentina			
	Nov. 01	08:51	22:23
	Nov. 30	08:34	22:51
Cape Town, South Africa			
	Nov. 01	03:45	17:14
	Nov. 30	03:28	17:42
London, UK			
	Nov. 01	06:55	16:33
	Nov. 30	07:44	15:55
Los Angeles, USA			
	Nov. 01	14:13	01:01
	Nov. 30	14:40	00:44
Nairobi, Kenya			
	Nov. 01	03:12	15:21
	Nov. 30	03:16	15:27
Sydney, Australia			
	Nov. 01	18:54	08:23
	Nov. 30	18:37	08:50
Tokyo, Japan			
	Nov. 01	21:04	07:46
	Nov. 30	21:32	07:28
Washington, DC, USA			
	Nov. 01	11:36	22:07
	Nov. 30	12:07	21:47
Wellington, New Zealand			
	Nov. 01	17:07	07:02
	Nov. 30	16:42	07:37

NB: *the times given are in Universal Time (UT)*

The Moon's phases and ages

Northern hemisphere

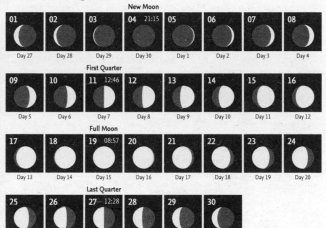

Southern hemisphere

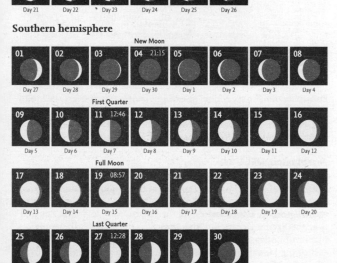

The Moon

On November 3, the extremely narrow waning crescent Moon
(it is one day before New Moon) passes close to *Mercury* and *Spica*
in *Virgo*, but this is too close to the Sun to be visible. On November
7–8, it may be possible to glimpse the young waxing crescent (Days
3 and 4 of the lunation), very low in the evening sky, when it is close
to *Venus*. On November 10, the Moon passes 4.1° south of *Saturn*
and then, the next day, on November 11, 4.4° south of *Jupiter* (both
planets are in *Capricornus*). On November 19, at Full Moon, there
is a partial lunar eclipse (described on the facing page). Just the
beginning of the eclipse is visible from western Europe, including
the UK, but it is fully visible from North and South America, the
Pacific, Australia and most of Asia. The next day (November 20) the
Moon is north of *Aldebaran* in *Taurus*. On November 24 it is 2.5°
south of *Pollux* in *Gemini*, and then on November 26, 2.5° south of
Regulus in *Leo*. By November 30 it is north of Spica in Virgo, as at
the beginning of the month, but again invisible, close to the Sun.

Beaver Moon
The November Full Moon has come to be called 'Beaver Moon',
because beavers become particularly active at this time, preparing
their lodges and food supplies for winter. But this applies to a
few only of the North-American tribes. Most see it as marking the
beginning of heavy frosts, with names such as 'Freezing River Maker
Moon', 'Freezing Moon', 'Rivers Begin to Freeze Moon', and 'Frost
Moon', although to the Haida of Alaska, where snow lies for many
months, it was 'Snow Moon'. In the Old World it was sometimes
known as the 'Frosty Moon' or even, occasionally, as the 'Oak Moon',
although the latter term was more often applied to the Full Moon
in December. If it was the last Full Moon before the winter solstice
it was also called the 'Mourning Moon'.

Partial lunar eclipse November 19
The second lunar eclipse of 2021 is technically classed as a partial
eclipse because a very small fraction of the Moon remains within
the penumbra, outside the central umbra. It is almost, but not quite,
a 'total' eclipse. This eclipse, like the one on May 26, is also centred
over the Pacific Ocean, but, this time, it is over the northern region.
Maximum eclipse (shown by the black dot) occurs at 09:27 UT to the
east of the Hawaiian Islands. The greatest phase, beginning at 07:18,
when the Moon starts to enter the Earth's umbra, will be visible

from the whole of North America, and last until 10:47 UT (the Moon will then be setting as seen from the West Coast of the United States and Canada). At that time, the eclipsed Moon will be rising over northern Australia, where only the final penumbral phase will be readily detectable. (But, of course, the penumbral phases of any lunar eclipse are effectively invisible to the naked eye, although the change in illumination may be detectable with carefully exposed photographs.)

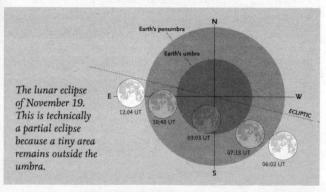

The lunar eclipse of November 19. This is technically a partial eclipse because a tiny area remains outside the umbra.

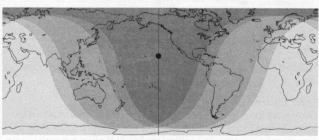

The lines on the map indicate the places where different phases of the eclipse begin. Starting in the east (right) they show when the Moon first contacts the penumbra; when it fully enters the penumbra; first contact with the umbra; and when the greatest area is within the dark umbra. The lines on the western side are the corresponding phases at the end of the eclipse.

N

Calendar for November

01	02:00 *	Mercury 4.4°N of Spica
03	11:49	Spica 5.8°S of Moon
03	18:39	Mercury 1.2°S of Moon
04	04:36	Mars 2.3°S of Moon
04	21:15	New Moon
04	23:58	Uranus at opposition (mag. 5.7)
05	22:18	Moon at perigee (358,844 km)
05–29		Leonid meteor shower
06	15:58	Antares 3.9°S of Moon
08	05:21	Venus 1.1°S of Moon
10	04:00 *	Mercury 1.1°N of Mars
10	14:24	Saturn 4.1°N of Moon
11	12:46	First Quarter
11	17:16	Jupiter 4.4°N of Moon
12		Northern Taurid shower maximum
13	18:37	Neptune 4.2°N of Moon
17–18		Leonid shower maximum
18	01:51	Uranus 1.5°N of Moon
19	08:57	Full Moon
19	09:27	Partial lunar eclipse (Americas, Pacific, Australia, Asia)
20	12:53	Aldebaran 6.4°S of Moon
21	02:13	Moon at apogee (406,279 km)
22–Dec.09		Phoenicid meteor shower
24	03:58	Pollux 2.5°N of Moon
26	23:00	Regulus 5.2°S of Moon
27	03:50	Dwarf planet (1) Ceres at opposition (mag. 7.0)
27	12:28	Last Quarter
29	04:39	Mercury at superior conjunction
30–Dec.14		Puppid Velid meteor shower
30	22:35	Spica 5.9°S of Moon

These objects are close together for an extended period around this time.

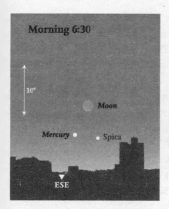

November 3 • *Mercury, Spica, and the crescent Moon form a nice triangle (as seen from London).*

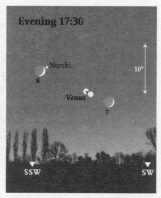

November 7–8 • *The waxing crescent Moon passes Venus and Nunki (σ Sgr), shortly after sunset (as seen from central USA).*

November 18–20 • *Venus passes Nunki (σ Sgr). Note that the scale of the illustration is different from the others (as seen from London).*

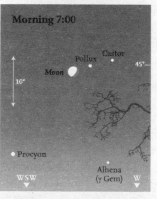

November 24 • *The Moon lines up with Castor and Pollux, high in the western sky (as seen from London).*

N

November – Looking North

For observers at mid-northern latitudes, the Milky Way is now high in the north and northwest. **Cassiopeia** is not far from the zenith and **Cepheus** is 'on its side' in the northwest. **Capella** (α Aurigae) and the whole constellation of **Auriga** are high in the northeast. One of the stars of the Summer Triangle, **Altair** in **Aquila**, is now disappearing below the horizon. The other two: **Deneb** in **Cygnus** and **Vega** in **Lyra** are still visible, but low towards the northwestern horizon. **Eltanin** (γ Draconis) the brightest star in the 'head' of **Draco**, is at about the same altitude as Vega. **Ursa Minor**, together with **Polaris** itself, is slightly higher in the sky.

The end of the 'tail' of **Ursa Major** (that is, the star **Alkaid**, η Ursae Majoris), is almost due north, although the body of the constellation has begun to swing round into the northeast. Most of the fainter stars in the constellation, to the south and east are easily visible, as is the undistinguished constellation of **Canes Venatici** to the southwest of it. Another insignificant constellation, often ignored, is **Leo Minor** to the south of Ursa Major. This consists of little more than three faint stars. **Gemini**, with the two bright stars, **Castor** and **Pollux**, is off to the east, with the line of stars forming **Lynx** between that constellation and Ursa Major. To the west of Ursa Major is **Hercules** although some of its stars are very low, close to the horizon, and likely to be difficult to see. Slightly towards the south are the northernmost stars of **Boötes**, although **Arcturus** itself is below the northern horizon.

*On 17 November 1970, the **Lunokhod 1** rover landed in western Mare Imbrium. Communications ceased on 14 September 1971.*

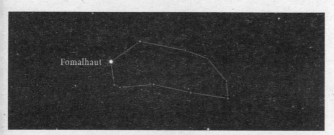

Fomalhaut

West of the meridian this month is the constellation of Piscis Austrinus, that has only one prominent star, the bright Fomalhaut.

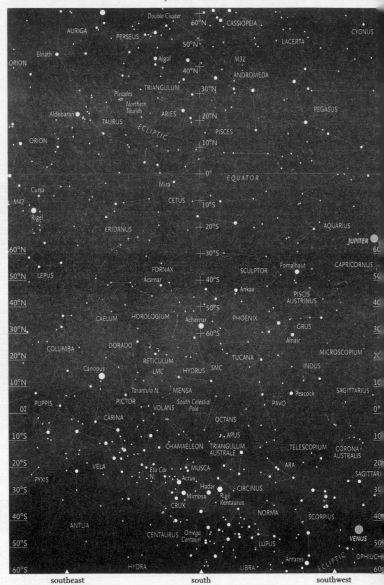

November – Looking South

Several major constellations dominate the southern sky. There is **Pegasus** with the Great Square and **Andromeda** above and to the east of it. **Pisces** straddles the meridian, with the 'Circlet' to the south of the Great Square. The zodiacal constellation of **Aquarius** is sinking in the west, but is still clearly visible. To the east is **Taurus** and orange-tinted **Aldebaran**. Below (south) of Pisces the whole of **Cetus** and its famous variable star, **Mira** (o Ceti), is clearly visible. **Orion** is rising in the east, and the beginning of **Eridanus** at the stars λ Eridani and **Cursa** (β Eridani) near **Rigel** (β Orionis) is visible as are most of the stars in the long chain that forms this constellation as it winds its way south. The brightest star, **Achernar** (α Eridani), at the very end of the constellation, is on the horizon for observers at a latitude of 30° north. This constellation once ended at **Acamar**, the moderately bright star south of the small constellation of **Fornax**, and originally called Achernar until the constellation was extended and the name transferred to the current, brighter star.

*The **Apollo 12** mission landed on the Moon on 19 November 1969, in Oceanus Procellarum (the Ocean of Storms).*

Immediately north and west of Achernar is **Phoenix** and between that and Aquarius lie the two constellations of **Sculptor** and **Piscis Austrinus**, the latter with its solitary bright star, **Fomalhaut**. Below Piscis Austrinus is the constellation of **Grus**. Both the Magellanic Clouds are clearly seen. The Small Cloud **(SMC)** on the border between **Tucana** and **Hydrus** is almost on the meridian, and the Large Cloud **(LMC)** is farther east, bordered by **Dorado** and **Mensa**. **Canopus** (α Carinae) is at almost the same altitude, and the stars of **Puppis** are farther east. South of Canopus are the stars of **Vela** and **Carina**. **Crux** is now more-or-less 'upright', just east of the meridian. **Rigil Kentaurus** and **Hadar** are on the other side of the meridian. All these significant stars are right on the horizon for observers

*On 26 November 2011, the NASA rover **Curiosity** landed at Gale crater on Mars.*

at 30° south, and only visible later in the night and later in the month. The zodiacal constellation of **Scorpius** is disappearing in the southwest and the stars of the sprawling constellation of **Centaurus** and those of **Lupus** are low in the sky.

N

December

December – Introduction

On December 4 there is another solar eclipse. Again in the polar regions, like the one on June 10 (pages 122–123), but this time over Antarctica, rather than over the Arctic. It is also a total eclipse, not an annular one, and occurs very close to perigee (which comes about two-and-a-half hours later and is actually the closest of the year). Full details of the eclipse and a diagram showing the path of the shadow are given on pages 214–215.

The last major meteor shower of the year is the *Geminids*. This reliable shower begins on December 3, and continues until December 16. The peak comes on December 14, when a rate of about 100 meteors per hour may be expected, with significant activity before midnight. Geminid meteors are slower than most other meteors and seem to last longer. They consist of denser material than normal (cometary) particles and are believed to come from the minor planet *(3200) Phaethon*, which has a similar orbit.

On 2 December 1971, the Soviet *Mars 3* probe was the first to make a soft landing on Mars. Unfortunately, radio transmissions failed after just 14.5 seconds.

The southern shower of the *Phoenicids* began on November 22, peaks on December 2, and ends on December 9. This shower is a bit of a mystery, partly because it is believed to be associated with Comet D/1819 W1 (Blanpain). This comet has dissipated (as indicated by the 'D') and there are no indications of the location of the remnants, so estimates of the possible rate are no more than guesswork. It is known, however, that there are often bright meteors, and that they tend to be slow. There is a second minor, southern shower, the *Puppid Velids*, that began on November 30, lasts

On 6 December 2014, the *New Horizons* spaceprobe was 'woken' from its hibernation state in preparation for its fly-by of the dwarf planet Pluto.

until December 14, and peaks on December 7. This has a predicted low maximum rate of about 10 meteors per hour. Finally, there is a minor northern shower, the *Ursids*. This begins on December 17, continues until December 26, with maximum on the night of December 22–23. The maximum rate is expected to be less than 10 meteors per hour. The parent body is Comet 8P/Tuttle.

The Planets

Mercury remains too close to the Sun to be readily visible. It was actually occulted by the Moon on November 3. Towards the end of the month, it moves into evening twilight and, with luck, may be glimpsed low in the sky, close to Venus. **Venus** itself is in **Sagittarius**, but very close to the Sun. Again, it may be glimpsed for a very short time in the evening sky at the end of the month. **Mars** begins the month in **Libra**, but soon becomes lost as it approaches the Sun. **Jupiter** (mag. -2.3 to -2.1) is in **Capricornus**, moving eastwards (with direct motion) and is visible in the evening sky. **Saturn** (mag. 0.7) is also in Capricornus, with slow direct motion, but closer to the horizon than Jupiter. **Uranus** (mag. 5.7) is slowly retrograding in **Aries. Neptune** (mag. 7.8) is in **Aquarius**, and resumes direct motion on December 4.

This time of the year, the constellation of Cetus and its famous variable star, Mira (o Ceti), is clearly visible. Because the celestial equator crosses the constellation, it is visible from almost every latitude.

D

Sunrise and sunset

City	Date	Sunrise	Sunset
Buenos Aires, Argentina			
	Dec. 01	08:34	22:52
	Dec. 31	08:43	23:10
Cape Town, South Africa			
	Dec. 01	03:28	17:43
	Dec. 31	03:38	18:00
London, UK			
	Dec. 01	07:45	15:55
	Dec. 31	08:07	16:01
Los Angeles, USA			
	Dec. 01	14:41	00:44
	Dec. 31	14:59	00:53
Nairobi, Kenya			
	Dec. 01	03:16	15:28
	Dec. 31	03:30	15:42
Sydney, Australia			
	Dec. 01	18:37	08:51
	Dec. 31	18:47	09:09
Tokyo, Japan			
	Dec. 01	21:33	07:28
	Dec. 31	21:51	07:37
Washington, DC, USA			
	Dec. 01	12:08	21:46
	Dec. 31	12:27	21:56
Wellington, New Zealand			
	Dec. 01	16:42	07:38
	Dec. 31	16:51	07:57

NB: the times given are in Universal Time (UT)

The Moon's phases and ages

Northern hemisphere

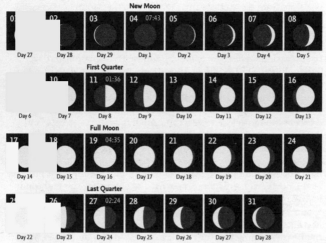

Southern hemisphere

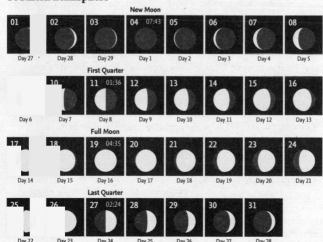

The Moon

At New Moon on December 4, there is a total solar eclipse, but this is visible from Antarctica only. On that day, the Moon is at perigee, at a distance of 356,794 km, which happens to be the closest it is to the Earth in 2021. It is actually 3.9° north of **Antares** in **Scorpius**. In the evening sky on December 7 to 9, the young waxing Moon passes, in succession, **Venus** (December 7), **Saturn** (December 8) and **Jupiter** (December 9). On December 17, waxing gibbous, two days before Full, the Moon is 6.4° north of **Aldebaran**. On December 21, now waning gibbous, it is 3.0° south of **Pollux** in **Gemini**. On December 24 it passes 5.1° north of **Regulus**. On December 31, the waning crescent Moon is close to **Mars** and **Antares**, and passes between the two.

Cold Moon
Because the cold of winter begins to dominate life for most places in the northern hemisphere, many of the names for the Full Moon in December refer to the temperature, with term such as 'Cold Moon', 'Winter Maker Moon' and 'Snow Moon'. The Cree of Canada had a rather strange name: 'Moon when the Young Fellow Spreads the Brush'. Among the Zuñi of New Mexico it was the 'Moon when the Sun has travelled home to rest'. In Europe it was sometimes called the 'Moon before Yule' or the 'Wolf Moon', although that term was more commonly applied to the Full Moon in January.

Total eclipse
The annular eclipse of June 10 occurs in the Arctic, but the next, total solar eclipse of December 4, occurs at the other end of the world, across Antarctica. Even the partial phases are visible from a far smaller area of land than with the eclipse six months earlier. On December 4, the partial phases are mainly visible from the southern Atlantic and Pacific Oceans, with the final phases seen from southeastern Australia. The very beginning (and the smallest possible 'bite' taken from the Sun) is visible from a tiny area of Namibia. ('Look away for a moment, and you will miss it!')

On 8 December 1995, NASA's *Galileo* spaceprobe became the first artificial satellite of Jupiter. It continued to function until 21 September 2003, when it was deliberately de-orbited to plunge into the upper atmosphere and be destroyed.

Maximum eclipse occurs at 07:33 UT and is as far south as 76°47' south, over the southern Weddell Sea near the Ronne Ice Shelf. The closest, inhabited research base is the Argentinian base of Belgrano II, just outside the area of totality. The British Halley VI Base is slightly farther away from the central line. At maximum eclipse the Sun will be covered for just 1 minute, 54.4 seconds. New Moon occurs slightly more than three hours later, at 07:43 UT, and the Moon is at perigee, which happens to be the closest during 2021, at 10:04 UT. The Moon's distance at perigee (when it is moving fastest) will be 356,794 km.

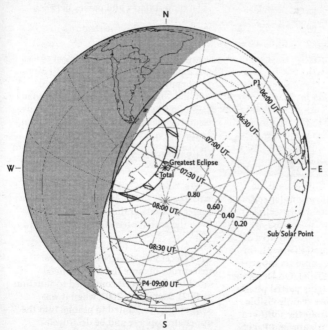

The total eclipse of December 4. The path of totality crosses East Antarctica, coming to maximum just offshore in the south of the Weddell Sea. The partial phases are visible over a wide area of Antarctica and the Southern Ocean.

D

Calendar for December

02		Phoenicid shower maximum
03–16		Geminid meteor shower
03	00:28	Mars 0.7°S of Moon
04	03:09	Antares 3.9°S of Moon
04	07:33	Total solar eclipse (Antarctica)
04	07:43	New Moon
04	10:04	Moon at perigee (356,794 km, closest of year)
04	12:43	Mercury 0.0°N of Moon
07		Puppid Velid shower maximum
07	00:48	Venus 1.98°N of Moon
08	01:49	Saturn 4.2°N of Moon
09	06:10	Jupiter 4.5°N of Moon
11	00:44	Neptune 4.2°N of Moon
11	01:36	First Quarter
14		Geminid shower maximum
15	05:53	Uranus 1.5°N of Moon
17–26		Ursid meteor shower
17	19:05	Aldebaran 6.4°S of Moon
18	02:15	Moon at apogee (406,320 km)
19	04:35	Full Moon
21	09:56	Pollux 3.0°N of Moon
21	15:59	December solstice
22–23		Ursid shower maximum
24	05:14	Regulus 5.1°S of Moon
26	18:00 *	Mars 4.6°N of Antares
27	02:24	Last Quarter
28	07:30	Spica 5.8°S of Moon
29	01:00 *	Mercury 4.2°S of Venus
31	14:24	Antares 3.9°S of Moon
31	20:13	Mars 1.0°N of Moon

These objects are close together for an extended period around this time.

December 6–9 • *The waxing crescent Moon passes Venus and Saturn and Jupiter in the south-southwestern sky (as seen from London).*

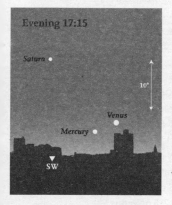

December 29 • *Mercury, Venus and Saturn in the southwest, after sunset (as seen from central USA).*

December 31 • *The crescent Moon with Antares, Mars and Sabik (η Oph) in an almost straight line (as seen from central USA).*

D

217

December – Looking North

For observers close to the equator, **Orion** is high overhead, to the east of the meridian, with **Taurus**, **Auriga** and **Gemini** way above the northern horizon. For observers at mid-northern latitudes, it is **Perseus** that is at the zenith, with **Andromeda** stretching off to the west and the Great Square of **Pegasus** in the northwest. **Auriga** with **Capella** is slightly to the east. Even farther east are the two bright stars of **Gemini**, **Castor** and **Pollux**.

The Milky Way runs from high in the northeast down to the northwest. Below Perseus is the distinctive shape of **Cassiopeia**, which lies within the Milky Way, and farther down is the zig-zag constellation of **Lacerta**, which is like **Cepheus**, in that both of them are partly within the band of stars. Even farther towards the northwest is **Cygnus** and the beginning of the Great Dark Rift near **Deneb**. The constellation of **Lyra**, with brilliant **Vega**, lies towards the meridian, away from the star clouds of the Milky Way. Much of the constellation of **Hercules** is clear of the northern horizon, together with some of the northernmost stars in **Boötes**.

Ursa Major is climbing in the northeast, and all the far-flung outlying stars are clearly visible. Below the 'tail' is the inconspicuous constellation of **Canes Venatici**, and some observers at high latitudes may even be able to glimpse some of the stars of **Coma Berenices**, low on the horizon in the northeast. Between Ursa Major and the Milky Way the whole of the constellations of **Draco** and **Ursa Minor** are clearly visible as is Cepheus beyond them. The chain of stars forming the faint constellation of **Lynx** lies between Ursa Major and the bright stars of Gemini.

On 24 December 1968, **Apollo 8** became the first crewed spacecraft to orbit the Moon, returning to Earth on December 27.

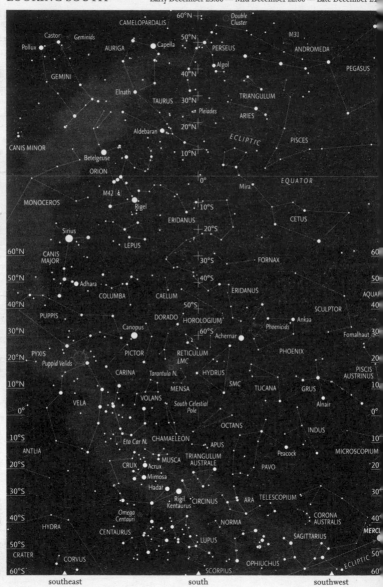

southeast   south   southwest

December – Looking South

Orion has now risen well clear of the horizon and is in the northeast. Above it are **Taurus** and **Auriga**, with **Gemini** farther to the east. Within the Milky Way to the east of Orion, is the rather undistinguished constellation of **Monoceros**, which is without any distinct, bright stars. Below Orion is the small constellation of **Lepus**, and to its east, **Canis Major**, with brilliant **Sirius**, the brightest star in the sky (mag. -1.4). The long, winding constellation of **Eridanus** begins near **Rigel** in Orion and runs south, partly enclosing **Fornax** (which was once part of the larger constellation), until it ends at **Achernar** (α Eridani). Between Eridanus and Canis Major is the tiny, faint constellation of **Caelum** and the larger and brighter **Columba**.

South of Canis Major is the constellation of **Puppis**, once (with **Vela** and **Carina**) part of the large obsolete constellation of Argo Navis. **Canopus** (α Carinae) and Achernar are close to the horizon for observers at 30°N. West of Achernar is **Phoenix** and even farther west, below **Aquarius**, the constellation of **Piscis Austrinus** is slowly descending in the west. Between Piscis Austrinus in the west and Carina in the east lie several small constellations, the most conspicuous of which is **Grus**. There are also **Tucana** and **Hydrus**, with the **Small Magellanic Cloud** (SMC), then the **Large Magellanic Cloud** (LMC) in **Mensa** and **Dorado**, with the tiny constellation of **Reticulum** north of them. North of that again is the faint constellation of **Horologium**. Between Mensa and the Milky Way in Carina is the tiny constellation of **Volans**.

More faint constellations lie around the South Celestial Pole in **Octans**, notably **Chamaeleon** and **Apus**. To the west lies the larger **Pavo** and the rather undistinguished constellation of **Indus**. **Crux** has now swung round into the southeast. Both it and the small neighbouring constellation of **Musca** lie in the Milky Way. **Rigil Kentaurus** and **Hadar**, the two brightest stars in **Centaurus** are slightly farther south, with **Circinus** and **Triangulum Australe** between them and Pavo. Those in the far south are able to see all the stars in Centaurus and **Lupus**, together with the southernmost stars of **Scorpius** (the 'sting') and **Sagittarius**, as well as other constellations, such as **Norma**, **Ara**, **Telescopium** and **Corona Australis**.

D

Johannes Kepler

The famous astronomer Johannes Kepler was born on 27 December 1571, 450 years ago, in Weil der Stadt, now in Baden-Württemberg in Germany. He attended the University of Tübingen and, at age 23, became a mathematics teacher at a school in Graz in Austria. Here, he published his first astronomical work, the *Mysterium Cosmographicum* (*The Cosmographic Mystery*, of 1596), which was the first book to defend the Copernican heliocentric system. In 1600, Kepler met the famous Danish astronomer, Tycho Brahe, who had a new observatory near Prague, and whom he impressed with his mathematical abilities. Eventually Kepler moved to Prague. After Tycho's death on 24 October 1601, Kepler was appointed imperial mathematician to the emperor, Rudolph II, a post in which he spent the next 11 years of his life. After the death of the emperor, Kepler moved to Linz, where he published his most important work *Epitome Astronomiae Copernicanae* (*Epitome of Copernican Astronomy*) in 1617. This contained Kepler's three laws of planetary motion, for which he would subsequently become famous. In 1627 he predicted future transits of Mercury and Venus for 1631, although the methods he used were not sufficiently accurate for him to predict the locations from which transits would be visible. The transit of Mercury (page 70) was observed by Pierre Gassendi (see page 171). He died on 15 November 1630 in Regensburg, now in Bavaria in Germany.

Wilhelm Tempel

Wilhelm Tempel (1821–1889) discovered a total of 21 comets, including Comet 55P/Tempel-Tuttle, now known to be the parent body of the **Leonid** meteor shower, which is noted for the major showers that occur at intervals of 33 years and peaks on November 17. Another was Comet 9P/Tempel, the target of the NASA probe **Deep Impact** in 2005. He also discovered five minor planets. One minor planet (3808 Tempel) and a lunar crater, some 45 km across, are named after him.

On 24 December 19, the Soviet Union's **Luna 13** spaceprobe landed in Oceanus Procellarum and returned images and data until December 28.

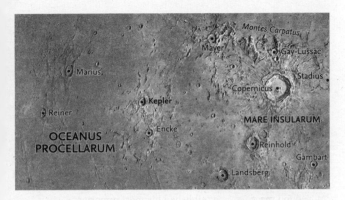

The lunar crater named after Johannes Kepler lies between Oceanus Procellarum and Mare Insularum. It is 32 km in diameter and is noted for being the source of a bright ray system, some rays of which extend for more than 300 km.

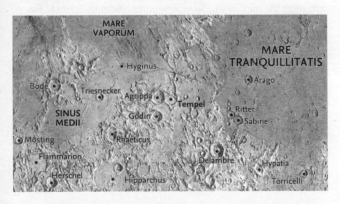

The lunar feature named after Wilhelm Tempel, is a greatly degraded remnant crater, whose walls have been broken by later impacts and eroded by major lava flows. The remnant is some 43 km across, located in the highland area between Sinus Medii and Mare Tranquillitatis. It is immediately adjacent to the distinctive crater Agrippa.

D

Dark Sky Sites

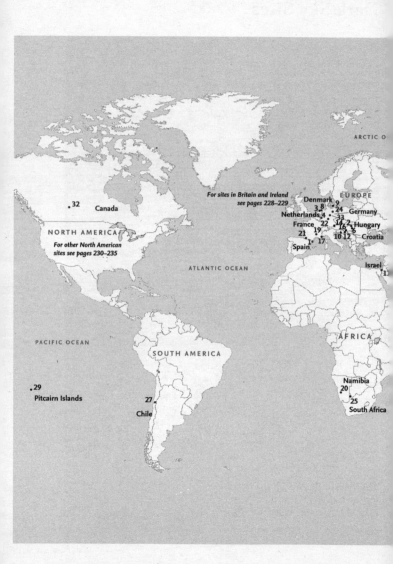

ARCTIC O

For sites in Britain and Ireland
see pages 228–229

EUROPE

Denmark
3 8 9
Netherlands 4 33 24 Germany
France 22 14 2 Hungary
19 16 6
21 10 12 Croatia
1 17
Spain

32 Canada

NORTH AMERICA
For other North American
sites see pages 230–235

ATLANTIC OCEAN

Israel
11

AFRICA

PACIFIC OCEAN

SOUTH AMERICA

Namibia
20

29
Pitcairn Islands

27
Chile

25
South Africa

ARCTIC OCEAN

OPE

ermany

Hungary

Croatia

Israel

11

ASIA

South
Korea

15

Japan

Taiwan 5

7

PACIFIC OCEAN

RICA

INDIAN OCEAN

mibia

25

outh Africa

OCEANIA

28
Niue

Australia

31

13

23

26

New Zealand

18

30

The 'Rest of the World' site details,
for the numbers shown on this
map, are given on pages 236–237.

British and Irish Dark Sky Sites

International Dark-Sky Association Sites

The International Dark-Sky Association (IDA) recognizes various categories of sites that offer areas where the sky is dark at night, free from light pollution and particularly suitable for astronomical observing. A number of sites in Great Britain and Ireland have been given specific recognition and are shown on the map. These are:

1   *Ballycroy National Park and Wild Nephin Wilderness*
2   *Bodmin Moor Dark Sky Landscape*
3   *Cranbourne Chase*
4   *Elan Valley Estate*
5   *Galloway Forest Park*
6   *Northumberland National Park and Kielder Water & Forest Park*
7   *Tomintoul and Glenlivet – Cairngorms*
8   *Brecon Beacons National Park*
9   *Exmoor National Park*
10 *Kerry*
11 *Moffat*
12 *Moore's Reserve – South Downs National Park*
13 *Snowdonia National Park*
14 *The island of Coll (Inner Hebrides, Scotland)*
15 *The island of Sark (Channel Islands)*

Details of these sites and web links may be found at the IDA website: https://www.darksky.org/
Many of these sites have major observatories or other facilities available for public observing (often at specific dates or times).

Dark Sky Discovery Sites

In Britain there is also the Dark Sky Discovery organisation. This gives recognition to smaller sites, again free from immediate light pollution, that are open to observing at any time. Some sites are used for specific, public observing sessions.
A full listing of sites is at:
https://www.darkskydiscovery.org.uk/
but specific events are publicized locally.

US Dark Sky Sites

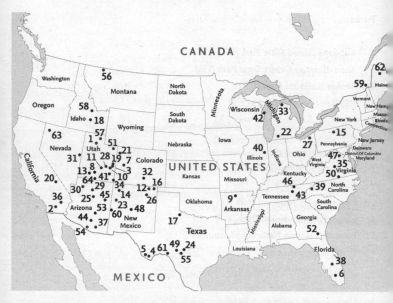

International Dark-Sky Association Sites

The International Dark-Sky Association (IDA) recognizes various categories of sites that offer areas where the sky is dark at night, free from light pollution and particularly suitable for astronomical observing. There are numerous sites in North America, shown on the map and listed here. Details of the IDA are at: https://www.darksky.org/.

Information on the various categories and individual sites are at: https://www.darksky.org/our-work/conservation/idsp/

Many of these sites have major observatories or other facilities available for public observing (often at specific dates or times).

Parks

1   *Antelope Island State Park (UT)*
2   *Anza-Borrego Desert State Park (CA)*
3   *Arches National Park (UT)*
4   *Big Bend National Park (TX)*
5   *Big Bend Ranch State Park (TX)*
6   *Big Cypress National Preserve (FL)*
7   *Black Canyon of the Gunnison National Park (CO)*
8   *Bryce Canyon National Park (UT)*
9   *Buffalo National River (AR)*
10 *Canyonlands National Park (UT)*
11 *Capitol Reef National Park (UT)*
12 *Capulin Volcano National Monument (NM)*
13 *Cedar Breaks National Monument (UT)*
14 *Chaco Culture National Historical Park (NM)*
15 *Cherry Springs State Park (PA)*
16 *Clayton Lake State Park (NM)*
17 *Copper Breaks State Park (TX)*
18 *Craters Of The Moon National Monument (ID)*
19 *Dead Horse Point State Park (UT)*
20 *Death Valley National Park (CA)*
21 *Dinosaur National Monument (CO)*
22 *Dr. T.K. Lawless County Park (MI)*
23 *El Morro National Monument (NM)*
24 *Enchanted Rock State Natural Area (TX)*
25 *Flagstaff Area National Monuments (AZ)*
26 *Fort Union National Monument (NM)*
27 *Geauga Observatory Park (OH)*
28 *Goblin Valley State Park (UT)*
29 *Grand Canyon National Park (AZ)*

30 *Grand Canyon-Parashant National Monument (AZ)*
31 *Great Basin National Park (NV)*
32 *Great Sand Dunes National Park and Preserve (CO)*
33 *Headlands (MI)*
34 *Hovenweep National Monument (UT)*
35 *James River State Park (VA)*
36 *Joshua Tree National Park (CA)*
37 *Kartchner Caverns State Park (AZ)*
38 *Kissimmee Prairie Preserve State Park (FL)*
39 *Mayland Earth to Sky Park & Bare Dark Sky Observatory (NC)*
40 *Middle Fork River Forest Preserve (IL)*
41 *Natural Bridges National Monument (UT)*
42 *Newport State Park (WI)*
43 *Obed Wild and Scenic River (TN)*
44 *Oracle State Park (AZ)*
45 *Petrified Forest National Park (AZ)*
46 *Pickett CCC Memorial State Park & Pogue Creek Canyon State Natural Area (TN)*
47 *Rappahannock County Park (VA)*
48 *Salinas Pueblo Missions National Monument (NM)*
49 *South Llano River State Park (TX)*
50 *Staunton River State Park (VA)*
51 *Steinaker State Park (UT)*
52 *Stephen C. Foster State Park (GA)*
53 *Tonto National Monument (AZ)*
54 *Tumacácori National Historical Park (AZ)*
55 *UBarU Camp and Retreat Center (TX)*
56 *Waterton-Glacier International Peace Park (Canada/MT)*
57 *Weber County North Fork Park (UT) Reserves*
58 *Central Idaho (ID)*
59 *Mont-Mégantic (Québec) Sanctuaries*

60 *Cosmic Campground (NM)*
61 *Devils River State Natural Area – Del Norte Unit (TX)*
62 *Katahdin Woods and Waters National Monument (ME)*
63 *Massacre Rim (NV)*
64 *Rainbow Bridge National Monument (UT)*

Dark Sky communities (Not on the map)

65 *Beverly Shores (IA)*
66 *Big Park / Village of Oak Creek (AZ)*
67 *Borrego Springs (CA)*
68 *Camp Verde (AZ)*
69 *Cottonwood (AZ)*
70 *Dripping Springs (TX)*
71 *Flagstaff (AZ)*
72 *Fountain Hills (AZ)*
73 *Fredericksburg (TX)*
74 *Helper (UT)*
75 *Homer Glen (IL)*
76 *Horseshoe Bay (TX)*
77 *Ketchum (ID)*
78 *Lakewood Village (TX)*
79 *Norwood (CO)*
80 *Sedona (AZ)*
81 *Thunder Mountain Pootseev Nightsky (AZ)*
82 *Torrey (UT)*
83 *Westcliffe and Silver Cliff (CO)*
84 *Wimberley Valley (TX)*

RASC Recognized Dark-Sky Sites

Canadian Dark-Sky Sites

The Royal Astronomical Society of Canada (RASC) has developed formal guidelines and requirements for three types of light-restricted protected areas: Dark-Sky Preserves, Urban Star Parks and Nocturnal Preserves. The focus of the Canadian Program is primarily to protect the nocturnal environment; therefore, the outdoor lighting requirements are the most stringent, but also the most effective. Canadian Parks and other areas that meet these guidelines and successfully apply for one of these designations are officially recognized. Many parks across Canada have been designated in recent years – see the list below and the RASC website: https://www.rasc.ca/dark-sky-site-designations.

Dark-Sky Preserves

1   ***Torrance Barrens Dark-Sky Preserve*** (ON)

2   ***McDonald Park Dark-Sky Park*** (BC)

3   ***Cypress Hills Inter-Provincial Park Dark-Sky Preserve*** (SK/AB)

4   ***Point Pelee National Park*** (ON)

5   ***Beaver Hills and Elk Island National Park*** (AB)

6   ***Mont-Mégantic International Dark-Sky Preserve*** (QC)

7   *Gordon's Park* (ON)

8   *Grasslands National Park* (SK)

9   *Bruce Peninsula National Park* (ON)

10   *Kouchibouguac National Park* (NB)

11   *Mount Carleton Provincial Park* (NB)

12   *Kejimkujik National Park* (NS)

13   *Fundy National Park* (NB)

14   *Jasper National Park Dark-Sky Preserve* (AB)

15   *Bluewater Outdoor Education Centre – Wiarton* (ON)

16   *Wood Buffalo National Park* (AB)

17   *North Frontenac Township* (ON)

18   *Lakeland Provincial Park and Provincial Recreation Area* (AB)

19   *Killarney Provincial Park* (ON)

20   *Terra Nova National Park* (NL)

21   *Au Diable Vert* (QC)

22   *Lake Superior Provincial Park* (ON)

Urban Star Parks

23   *Irving Nature Park* (NB)

24   *Cattle Point, Victoria* (BC)

Nocturnal Preserves

25   *Ann and Sandy Cross Conservation Area* (AB)

26   *Old Man on His Back Ranch* (SK)

Dark Sky Parks – Rest of the World

1 *Albanyà (Spain)*

2 *Bükk National Park (Hungary)*

3 *De Boschplatt (Netherlands)*

4 *Eifel National Park (Germany)*

5 *Hehuan Mountain (Taiwan)*

6 *Hortobágy National Park (Hungary)*

7 *Iriomote-Ishigaki National Park (Japan)*

8 *Lauwersmeer National Park (Netherlands)*

9 *Møn and Nyord (Denmark)*

10 *Petrova gora-Biljeg (Croatia)*

11 *Ramon Crater (Israel)*

12 *Vrani kamen (Croatia)*

13 *Warrumbungle National Park (Australia)*

14 *Winklmoosalm (Germany)*

15 *Yeongyang Firefly Eco Park (South Korea)*

16 *Zselic National Landscape Protection Area (Hungary)*

Dark Sky Reserves

17 *Alpes Azur Mercantour (France)*

18 *Aoraki Mackenzie (New Zealand)*

19 *Cévennes National Park (France)*

20 *NambiRand Nature Reserve (Namibia)*

21 *Pic du Midi (France)*

22 *Rhön (Germany)*

23 *River Murray (Australia)*

24 *Westhavelland (Germany)*

Dark Sky Sanctuaries

25 *!Ae!Hai Kalahari Heritage Park (South Africa)*

26 *Aotea / Great Barrier Island (New Zealand)*

27 *Gabriela Mistral (Chile)*

28 *Niue (New Zealand)*

29 *Pitcairn Islands (UK)*

30 *Stewart Island / Rakiura (New Zealand)*

31 *The Jump-Up (Australia)*

Dark Sky Communities

32 *Bon Accord (Canada)*

33 *Fulda (Germany)*

Twilight Diagrams

Sunrise, sunset, twilight

For each individual month, we give details of sunrise and sunset times for nine cities across the world. But observing the stars is also affected by twilight, and this varies considerably from place to place. During the summer, especially at high latitudes, twilight may persist throughout the night and make it difficult to see the faintest stars. Beyond the Arctic and Antarctic Circles, of course, the Sun does not set for 24 hours at least once during the summer (and rise for 24 hours at least once during the winter). Even when the Sun does dip below the horizon at high latitudes, bright twilight persists throughout the night, so observing the stars is impossible.

There are three recognized stages of twilight: civil twilight, when the Sun is less than 6° below the horizon; nautical twilight, when the Sun is between 6° and 12° below the horizon; and astronomical twilight, when the Sun is between 12° and 18° below the horizon. Full darkness occurs only when the Sun is more than 18° below the horizon. During nautical twilight, only the very brightest stars are visible. (These are the stars that were used for navigation, hence the name for this stage.) During astronomical twilight, the faintest stars visible to the naked eye may be seen directly overhead, but are lost at lower altitudes. They become visible only once it is fully dark. The diagrams show the duration of twilight at the various cities. Of the locations shown, during the summer months there is full darkness at most of the cities, but it never occurs during the summer at the latitude of London. Observing conditions are most favourable at somewhere like Nairobi, which is very close to the equator, so there is not only little twilight, and a long period of full darkness, but there are also only slight variations in timing and duration throughout the year.

The diagrams also show the times of New and Full Moon (black and white symbols, respectively). As may be seen, at most locations during the year roughly half of New and Full Moon phases may come during daylight. For this reason, the exact phase may be invisible at one location, but be clearly seen elsewhere. The exact times of the events are given in the diagrams for each individual month.

■ Civil Twilight   ■ Nautical Twilight   ■ Astronomical Twilight   ■ Full Darkness

◇ Exact time of Full Moon   ◆ Exact time of New Moon

Buenos Aires, Argentina – Latitude 34.7°S – Longitude 58.5°W

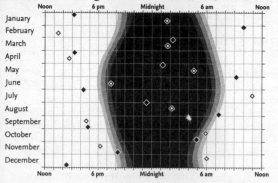

Legend on
page 238

Cape Town, South Africa – Latitude 33.9°S – Longitude 18.5°E

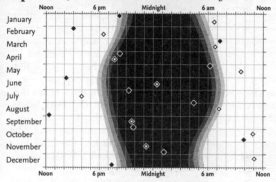

London, UK – Latitude 51.5°N – Longitude 2.0°W

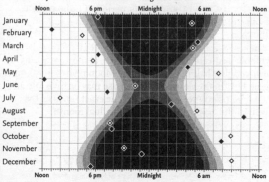

Los Angeles, USA – Latitude 34.0°N – Longitude 118.2° W

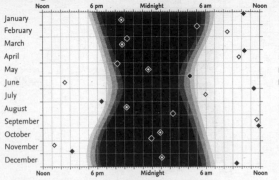

Legend on page 238

Nairobi, Kenya – Latitude 1.3°S – Longitude 36.8°E

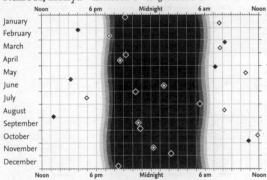

Sydney, Australia – Latitude 33.5°S – Longitude 151.2°E

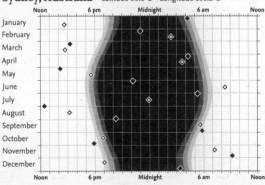

Tokyo, Japan – Latitude 35.7°N – Longitude 139.8°E

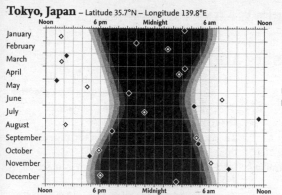

Legend on page 238

Washington, DC, USA – Latitude 38.9°N – Longitude 77.0°W

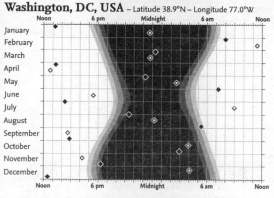

Wellington, New Zealand – Latitude 41.3°S – Longitude 174.8°E

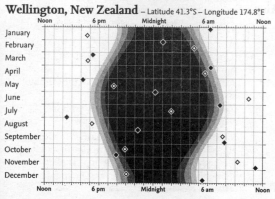

Glossary and Tables

aphelion	The point on an orbit that is farthest from the Sun.
apogee	The point on its orbit at which the Moon is farthest from the Earth.
appulse	The apparently close approach of two celestial objects; two planets, or a planet and star.
astronomical unit	(AU) The mean distance of the Earth from the Sun, 149,597,870 km.
celestial equator	The great circle on the celestial sphere that is in the same plane as the Earth's equator.
celestial sphere	The apparent sphere surrounding the Earth on which all celestial bodies (stars, planets, etc.) seem to be located.
conjunction	The point in time when two celestial objects have the same celestial longitude. In the case of the Sun and a planet, superior conjunction occurs when the planet lies on the far side of the Sun (as seen from Earth). For Mercury and Venus, inferior conjunction occurs when they pass between the Sun and the Earth.
direct motion	Motion from west to east on the sky.
ecliptic	The apparent path of the Sun across the sky throughout the year. Also: the plane of the Earth's orbit in space.
elongation	The point at which an inferior planet has the greatest angular distance from the Sun, as seen from Earth.
equinox	The two points during the year when night and day have equal duration. Also: the points on the sky at which the ecliptic intersects the celestial equator. The vernal (northern spring) equinox is of particular importance in astronomy.
gibbous	The stage in the sequence of phases at which the illumination of a body lies between half and full. In the case of the Moon, the term is applied to phases between First Quarter and Full, and between Full and Last Quarter.

inferior planet	Either of the planets Mercury or Venus, which have orbits inside that of the Earth.
magnitude	The brightness of a star, planet or other celestial body. It is a logarithmic scale, where larger numbers indicate fainter brightness. A difference of 5 in magnitude indicates a difference of 100 in actual brightness, thus a first-magnitude star is 100 times as bright as one of sixth magnitude.
meridian	The great circle passing through the North and South Poles of a body and the observer's position; or the corresponding great circle on the celestial sphere that passes through the North and South Celestial Poles and also through the observer's zenith.
nadir	The point on the celestial sphere directly beneath the observer's feet, opposite the zenith.
occultation	The disappearance of one celestial body behind another, such as when stars or planets are hidden behind the Moon.
opposition	The point on a superior planet's orbit at which it is directly opposite the Sun in the sky.
perigee	The point on its orbit at which the Moon is closest to the Earth.
perihelion	The point on an orbit that is closest to the Sun.
retrograde motion	Motion from east to west on the sky.
superior planet	A planet that has an orbit outside that of the Earth.
vernal equinox	The point at which the Sun, in its apparent motion along the ecliptic, crosses the celestial equator from south to north. Also known as the First Point of Aries.
zenith	The point directly above the observer's head.
zodiac	A band, stretching 8° on either side of the ecliptic, within which the Moon and planets appear to move. It consists of 12 equal areas, originally named after the constellation that once lay within it.

The Constellations

There are 88 constellations covering the whole of the celestial sphere. The names themselves are expressed in Latin, and the names of stars are frequently given by Greek letters (see page 246) followed by the genitive of the constellation name or its three-letter abbreviation. The genitives, the official abbreviations and the English names of the various constellations are included.

Name	Genitive	Abbr.	English name
Andromeda	Andromedae	And	Andromeda
Antlia	Antliae	Ant	Air Pump
Apus	Apodis	Aps	Bird of Paradise
Aquarius	Aquarii	Aqr	Water Bearer
Aquila	Aquilae	Aql	Eagle
Ara	Arae	Ara	Altar
Aries	Arietis	Ari	Ram
Auriga	Aurigae	Aur	Charioteer
Boötes	Boötis	Boo	Herdsman
Caelum	Caeli	Cae	Burin
Camelopardalis	Camelopardalis	Cam	Giraffe
Cancer	Cancri	Cnc	Crab
Canes Venatici	Canum Venaticorum	CVn	Hunting Dogs
Canis Major	Canis Majoris	CMa	Big Dog
Canis Minor	Canis Minoris	CMi	Little Dog
Capricornus	Capricorni	Cap	Sea Goat
Carina	Carinae	Car	Keel
Cassiopeia	Cassiopeiae	Cas	Cassiopeia
Centaurus	Centauri	Cen	Centaur
Cepheus	Cephei	Cep	Cepheus
Cetus	Ceti	Cet	Whale
Chamaeleon	Chamaeleontis	Cha	Chameleon
Circinus	Circini	Cir	Compasses
Columba	Columbae	Col	Dove
Coma Berenices	Comae Berenices	Com	Berenice's Hair
Corona Australis	Coronae Australis	CrA	Southern Crown
Corona Borealis	Coronae Borealis	CrB	Northern Crown
Corvus	Corvi	Crv	Crow

Name	Genitive	Abbr.	English name
Crater	Crateris	Crt	Cup
Crux	Crucis	Cru	Southern Cross
Cygnus	Cygni	Cyg	Swan
Delphinus	Delphini	Del	Dolphin
Dorado	Doradus	Dor	Dorado
Draco	Draconis	Dra	Dragon
Equuleus	Equulei	Equ	Little Horse
Eridanus	Eridani	Eri	River Eridanus
Fornax	Fornacis	For	Furnace
Gemini	Geminorum	Gem	Twins
Grus	Gruis	Gru	Crane
Hercules	Herculis	Her	Hercules
Horologium	Horologii	Hor	Clock
Hydra	Hydrae	Hya	Water Snake
Hydrus	Hydri	Hyi	Lesser Water Snake
Indus	Indi	Ind	Indian
Lacerta	Lacertae	Lac	Lizard
Leo	Leonis	Leo	Lion
Leo Minor	Leonis Minoris	LMi	Little Lion
Lepus	Leporis	Lep	Hare
Libra	Librae	Lib	Scales
Lupus	Lupi	Lup	Wolf
Lynx	Lyncis	Lyn	Lynx
Lyra	Lyrae	Lyr	Lyre
Mensa	Mensae	Men	Table Mountain
Microscopium	Microscopii	Mic	Microscope
Monoceros	Monocerotis	Mon	Unicorn
Musca	Muscae	Mus	Fly
Norma	Normae	Nor	Set Square
Octans	Octantis	Oct	Octant
Ophiuchus	Ophiuchi	Oph	Serpent Bearer
Orion	Orionis	Ori	Orion
Pavo	Pavonis	Pav	Peacock
Pegasus	Pegasi	Peg	Pegasus
Perseus	Persei	Per	Perseus
Phoenix	Phoenicis	Phe	Phoenix

Name	Genitive	Abbr.	English name
Pictor	Pictoris	Pic	Painter's Easel
Pisces	Piscium	Psc	Fishes
Piscis Austrinus	Piscis Austrini	PsA	Southern Fish
Puppis	Puppis	Pup	Stern
Pyxis	Pyxidis	Pyx	Compass
Reticulum	Reticuli	Ret	Net
Sagitta	Sagittae	Sge	Arrow
Sagittarius	Sagittarii	Sgr	Archer
Scorpius	Scorpii	Sco	Scorpion
Sculptor	Sulptoris	Scu	Sculptor
Scutum	Scuti	Sct	Shield
Serpens	Serpentis	Ser	Serpent
Sextans	Sextantis	Sex	Sextant
Taurus	Tauri	Tau	Bull
Telescopium	Telescopii	Tel	Telescope
Triangulum	Trianguli	Tri	Triangle
Triangulum Australe	Trianguli Australis	TrA	Southern Triangle
Tucana	Tucanae	Tuc	Toucan
Ursa Major	Ursae Majoris	UMa	Great Bear
Ursa Minor	Ursae Minoris	UMi	Lesser Bear
Vela	Velorum	Vel	Sails
Virgo	Virginis	Vir	Virgin
Volans	Volantis	Vol	Flying Fish
Vulpecula	Vulpeculae	Vul	Fox

The Greek Alphabet

α	Alpha	ι	Iota	ρ	Rho
β	Beta	κ	Kappa	σ (ς)	Sigma
γ	Gamma	λ	Lambda	τ	Tau
δ	Delta	μ	Mu	υ	Upsilon
ε	Epsilon	ν	Nu	φ (φ)	Phi
ζ	Zeta	ξ	Xi	χ	Chi
η	Eta	o	Omicron	ψ	Psi
θ (ϑ)	Theta	π	Pi	ω	Omega

Asterisms

Apart from the constellations (88 of which cover the whole sky), listed on pages 244–246, certain groups of stars, which may form a small part of a larger constellation, are readily recognizable and have been given individual names. These groups are known as *asterisms*, and the most famous (and well-known) is the 'Plough' or 'Big Dipper', the common name for the seven brightest stars in the constellation of Ursa Major, the Great Bear. The names and details of some asterisms mentioned in this book are given in this list.

Some common asterisms

Belt of Orion	δ, ε and ζ Orionis
Big Dipper	α, β, γ, δ, ε, ζ and η Ursae Majoris
Cat's Eyes	λ and υ Scorpii
Circlet	γ, θ, ι, λ and κ Piscium
False Cross	ε and ι Carinae and δ and κ Velorum
Fish Hook	α, β, δ and π Scorpii
Guards (or Guardians)	β and γ Ursae Minoris
Head of Cetus	α, γ, ξ², μ and λ Ceti
Head of Draco	β, γ, ξ and ν Draconis
Head of Hydra	δ, ε, ζ, η, ρ and σ Hydrae
Job's Coffin	α, β, γ and δ Delphini
Keystone	ε, ζ, η and π Herculis
Kids	ε, ζ and η Aurigae
Little Dipper	β, γ, η, ζ, ε, δ and α Ursae Minoris
Lozenge	= Head of Draco
Milk Dipper	ζ, γ, σ, φ and λ Sagittarii
Plough	= Big Dipper
Pointers	α and β Ursae Majoris
Pot	= Saucepan
Saucepan	ι, θ, ζ, ε, δ and η Orionis
Sickle	α, η, γ, ζ, μ and ε Leonis
Southern Pointers	α and β Centauri
Square of Pegasus	α, β and γ Pegasi with α Andromedae
Sword of Orion	θ and ι Orionis
Teapot	γ, ε, δ, λ, φ, σ, τ and ζ Sagittarii
Wain (or Charles' Wain)	= Big Dipper
Water Jar	γ, η, κ and ζ Aquarii
Y of Aquarius	= Water Jar

Further Information

Books

Bone, Neil (1993), *Observer's Handbook: Meteors*, George Philip, London
& Sky Publishing Corp., Cambridge, Mass.

Cook, J., ed. (1999), *The Hatfield Photographic Lunar Atlas*, Springer-Verlag,
New York

Dunlop, Storm (2006), *Wild Guide to the Night Sky*, Harper Perennial, New York
& Smithsonian Press, Washington DC.

Dunlop, Storm (2012), *Practical Astronomy*, 2nd edn, Firefly, Buffalo

Dunlop, Storm, Rükl, Antonin & Tirion, Wil (2005), *Collins Atlas of the Night Sky*,
HarperCollins, London & Smithsonian Press, Washington DC.

Grego, Peter (2016), *Moon Observer's Guide*, Firefly, Richmond Hill

Heifetz, Milton D. & Tirion, Wil (2017), *A Walk through the Heavens*, 4th edn,
Cambridge University Press, Cambridge

Mellinger, Axel & Hoffmann, Susanne (2005), *The New Atlas of the Stars*,
Firefly, Richmond Hill

O'Meara, Stephen J. (2008), *Observing the Night Sky with Binoculars*,
Cambridge University Press, Cambridge

Pasachoff, Jay M. (1999), *Peterson Field Guides: Stars and Planets*, 4th edn.,
Houghton Mifflin, Boston

Ridpath, Ian, ed. (2003), *Oxford Dictionary of Astronomy*, 2nd edn, Oxford
University Press, Oxford & New York

Ridpath, Ian, ed. (2004), *Norton's Star Atlas*, 20th edn, Pi Press, New York

Ridpath, Ian (2018), *Star Tales*, 2nd Edn, Lutterworth Press, Cambridge UK

Ridpath, Ian & Tirion, Wil (2004), *Collins Gem – Stars*, HarperCollins, London

Ridpath, Ian & Tirion, Wil (2011), *Collins Pocket Guide Stars and Planets*, 4th edn,
HarperCollins, London

Ridpath, Ian & Tirion, Wil (2012), *Monthly Sky Guide*, 10th edn, Dover
Publications, New York

Rükl, Antonín (1990), *Hamlyn Atlas of the Moon*, Hamlyn, London
& Astro Media Inc., Milwaukee

Rükl, Antonín (2004), *Atlas of the Moon*, Sky Publishing Corp., Cambridge, Mass.

Scagell, Robin (2014), *Stargazing with a Telescope*, Firefly, Richmond Hill

Scagell, Robin (2015), *Firefly Complete Guide to Stargazing*, Firefly,
Richmond Hill

Scagell, Robin & Frydman, David (2014), *Stargazing with Binoculars*, Firefly,
Richmond Hill

Sky & Telescope (2017), *Astronomy 2018*, Sky Publishing Corp., Cambridge, Mass.

Stimac, Valerie (2019), *Dark Skies: A Practical Guide to Astrotourism*,
Lonely Planet

Tirion, Wil (2011), *Cambridge Star Atlas*, 4th edn, Cambridge University Press,
Cambridge

Tirion, Wil & Sinnott, Roger (1999), *Sky Atlas 2000.0*, 2nd edn, Sky Publishing
Corp., Cambridge, Mass. & Cambridge University Press, Cambridge

Journals

Astronomy, Astro Media Corp., 21027 Crossroads Circle, P.O. Box 1612,
Waukesha, WI 53187-1612.
http://www.astronomy.com

Sky & Telescope, Sky Publishing Corp., Cambridge, MA 02138-1200.
http://www.skyandtelescope.com/

Societies

American Association of Variable Star Observers (AAVSO), 49 Bay State Rd.,
Cambridge, MA 02138. Although primarily concerned with variable
stars, the AAVSO also has a solar section.

American Astronomical Society (AAS), 1667 K Street NW, Suite 800, Washington,
DC 20006, New York.
http://aas.org/

American Meteor Society (AMS), Geneseo, New York.
http://www.amsmeteors.org/

Association of Lunar and Planetary Observers (ALPO), ALPO Membership
Secretary/Treasurer, P.O. Box 13456, Springfield, IL 62791-3456.
An organization concerned with all forms of amateur astronomical
observation, not just the Moon and planets, with numerous
coordinated observing sections.
http://alpo-astronomy.org/

Astronomical League (AL), 9201 Ward Parkway Suite #100,
Kansas City, MO 64114.
An umbrella organization consisting of over 240 local amateur astronomical societies across the United States.
https://www.astroleague.org/

British Astronomical Association (BAA), Burlington House, Piccadilly, London W1J 0DU.
The principal British organization (but with a worldwide membership) for amateur astronomers (with some professional members), particularly for those interested in carrying out observational programs.
http://www.britastro.org/

International Meteor Organization (IMO)
An organization coordinating observations of meteors worldwide.
http://www.imo.net/

Royal Astronomical Society of Canada (RASC), 203 – 4920 Dundas Street W., Toronto, ON M9A 1B7.
The principal Canadian astronomical organization, with both professional and amateur members. It has 28 local centres.
http://rasc.ca/

Software

Planetary, Stellar and Lunar Visibility (planetary and eclipse freeware):
Alcyone Software, Germany.
http://www.alcyone.de

Redshift, Redshift-Live.
http://www.redshift-live.com/en/

Starry Night & Starry Night Pro, Sienna Software Inc., Toronto, Canada.
http://www.starrynight.com

Internet sources

There are numerous sites about all aspects of astronomy, and all have numerous links. Although many amateur sites are excellent, treat any statements and data with caution. The sites listed below offer accurate information. Please note that the URLs may change. If so, use a good search engine, such as Google, to locate the information source.

Information

Astronomical data (inc. eclipses) HM Nautical Almanac Office:
http://astro.ukho.gov.uk

Auroral information Michigan Tech:
http://www.geo.mtu.edu/weather/aurora/

Comets JPL Solar System Dynamics:
http://ssd.jpl.nasa.gov/

Deep-sky objects Saguaro Astronomy Club Database:
http://www.virtualcolony.com/sac/

Eclipses NASA Eclipse Page:
http://eclipse.gsfc.nasa.gov/eclipse.html
https://en.wikipedia.org/wiki/Solar_eclipse_of_June_10,_2021

Moon (inc. Atlas) Inconstant Moon:
http://www.inconstantmoon.com/

Planets Planetary Fact Sheets:
http://nssdc.gsfc.nasa.gov/planetary/planetfact.html

Satellites (inc. International Space Station)
Heavens Above: http://www.heavens-above.com/
Visual Satellite Observer: http://www.satobs.org/

Star Chart
http://www.skyandtelescope.com/observing/interactive-sky-watching-tools/interactive-sky-chart/

What's Visible
Skyhound: http://www.skyhound.com/sh/skyhound.html
Skyview Cafe: http://www.skyviewcafe.com

Institutes and Organizations

European Space Agency: http://www.esa.int/
International Dark-Sky Association: http://www.darksky.org/
RASC Dark Sky: https://www.rasc.ca/dark-sky-site-designations
Jet Propulsion Laboratory: http://www.jpl.nasa.gov/
Lunar and Planetary Institute: http://www.lpi.usra.edu/
National Aeronautics and Space Administration: http://www.hq.nasa.gov/
Solar Data Analysis Center: http://umbra.gsfc.nasa.gov/
Space Telescope Science Institute: http://www.stsci.edu/

Acknowledgements

5	Tom Kerss
21	NASA/GSFC/SVS
44	Stephen Pitt
52	Denis Buczynski
53	Ian Ridpath
63	NASA (Big Bertha)
70	Wikimedia Commons/NASA
76	NASA/JPL-Caltech/UCLA/MPS/DLR/IDA, (Ceres and Vesta) ESO (Pallas)
77	NASA
81	Mike Young
95	Duncan Waldron, Brisbane, Australia
102	Occult4 computer program by David Herald (map)
109	NASA/GSFC/Fred Espenak
113	Bernhard Hubl
123	NASA/GSFC/Fred Espenak
131	Alan Tough, Elgin, Scotland
139	NASA
146	Peter Komka/EPA-EFE/Shutterstock
150	Jens Hackmann
155	NASA/GSFC/Arizona State University (bottom)
163	NASA/ESA Hubble Space Telescope
171	NASA Lunar Orbiter 4
185	Duncan Waldron, Brisbane, Australia
193	Alan Tough, Elgin, Scotland
201	NASA/GSFC/Fred Espenak
215	NASA/GSFC/Fred Espenak
225	Bernhard Hubl
49, 57, 85, 99, 101, 105, 119, 159, 175, 181, 189, 205, 211	Based on photographs by Bernhard Hubl

The authors would also like to thank Barry Hetherington.

Specialist editorial support was provided by Hannah Banyard and Dr Gregory Brown, Public Astronomy Officers at the Royal Observatory, part of Royal Museums Greenwich.

Index

EXPLORE OUR RANGE OF ASTRONOMY TITLES

Other titles by Storm Dunlop and Wil Tirion

2021 Guide to the Night Sky: Britain and Ireland
978-0-00-838904-8

2021 Guide to the Night Sky: North America
978-0-00-839977-1

2021 Guide to the Night Sky: Southern Hemisphere
978-0-00-839979-5

Latest editions of our bestselling month-by-month guides for exploring the skies. These guides are an easy introduction to astronomy and a useful reference for seasoned stargazers.

Collins Planisphere | 978-0-00-754075-4
Easy-to-use practical tool to help astronomers to identify the constellations and stars every day of the year. For latitude 50°N, suitable for use anywhere in Britain and Ireland, Northern Europe, Canada and Northern USA

Also available

Astronomy Photographer of the Year: Collection 9
978-0-00-840463-5

Winning and shortlisted images from the 2020 Insight Investment Astronomy Photographer of the Year competition, hosted by the Royal Observatory, Greenwich. The images include aurorae, galaxies, our Moon, our Sun, people and space, planets, comets and asteroids, skyscapes, stars and nebulae.

Stargazing | 978-0-00-819627-1
The prefect manual for beginners to astronomy - introducing the world of telescopes, planets, stars, dark skies and celestial maps.

Moongazing | 978-0-00-830500-0
An in-depth guide for all aspiring astronomers and Moon observers, with detailed Moon maps. Covers the history of lunar exploration and the properties of the Moon, its origin and orbit.

The Moon | 978-0-00-828246-2
A celebration of our celestial neighbour exploring people's fascination with our only natural satellite, illuminating how art and science meet in our profound connection with the Moon.